ATILLO
BALINTAWAK ESKRIMA
THE ORIGINAL SAAVEDRA STYLE

Crispulo "Ising" Atillo
with Glen Boodry

Disclaimer

The authors of this book are not responsible whatsoever for any injuries that may result from practicing the techniques and/or following the instructions given within. Please consult a physician before training or engaging in any of the physical activities described in this book.

Published in 2020 by Crispulo Atillo and Glen Boodry

Copyright © 2020 Crispulo Atillo and Glen Boodry

All rights reserved. No part of this publication may be reproduced or utilized in any form or by any means, electronic or mechanical, including photocopying, recording, or any information storage or retrieval system, without prior written permission from the authors.

Library of Congress Control Number: 2020920002
ISBN: 978-0-578-78115-0

Dedication

This book is dedicated to the memory of my dad, Vicente "Inting" Atillo, my godfather, Delfin Lopez, and Teodoro "Doring" Saavedra. If not for their never-ending devotion to the practice of Cebuano eskrima and willingness to share their knowledge with me, the Atillo Balintawak Eskrima - Original Saavedra Style would not exist.

Balintawak Self Defense Club 1952
Cebu City, Philippines

For my wife, Beatriz "Betty" Atillo

In Loving Memory of my Mom
Faith Ellen Boodry
December 4, 1941 – August 20, 2019

Table of Contents

Forward	1
Preface	3
Acknowledgments	5
Letter of Endorsement .... Dan Inosanto	9
Letter of Endorsement .... Ronald Goldstein	10
Prologue	11
Introduction	12

CHAPTER 1: Lorenzo Saavedra 17

National Uprising and Call for Independence................	17
Rebel Accusation and Incarceration	20
The French School of Fencing	21
Development and Refinement During Incarceration	23

CHAPTER 2: Labangon 27

Labangon Fencing Club	28
Translation and Interpretation	30
Styles and Methods	30
Quarreling and Disagreements	32
Formal Disbandment	32

CHAPTER 3: The Doce Pares Club 35

Membership Growth and Controversy	37
Teodoro Saavedra vs. Pablo Alicante	38
Teodoro Saavedra vs. Pedrito Romo	41
Teodoro Saavedra vs. Roman Ladaño	42
Growth and Collaboration	42

CHAPTER 4: World War II ……………………..……………....	45
Japanese Invasion of the Philippines …………..…………....	46
Surrender of Bataan ……………………..……………….….	51
Japanese Invasion of Cebu ………..…………………….……..	52
Capture of Corregidor and Surrender to the Japanese ……...	54
James Cushing and the Cebuano Guerillas ……………….…	57
Assassination Mission ………………………………………..	62
Battle of Babag ………………………………………………..	63
Train Assault at Inayawan Crossing ………………………..	64
Farmhouse Ambush …………………………………………..	66
Death of an Eskrima Icon ……………………………………..	66
Capture and Escape from Death ……………………………...	69
Guerilla Recognition and the Koga Papers ……………….…	73
Recovery of the Z Plan Documents ……………………….….	76
Prisoner Arrival at Tabunan ……………………………….…	77
Death of Governor Hilario Abellana ………………………..	83
MacArthur's Return ……………………………………….…..	84
Liberation of Cebu ……………………………………….…....	85
Japanese Surrender ……………………………………….…..	88
Passing of an Eskrima Legend ……………………..………..	90
The Aftermath of War ……………………………….………..	92
CHAPTER 5: Balintawak ……………………………...……..…..	97
Internal Bickering and Politics ……………………….……...	97
Venancio "Anciong" Bacon ……………………………….…..	98
Venancio Bacon vs. Delfin Lopez ………………….………...	101
The Balintawak Self Defense Club ……………….………...	103
Styles and Systems ………………………………….………...	105

Beginning of a Legendary Rivalry 107

CHAPTER 6: Turbulent Era 109
    Cebu Politics and Delfin Lopez 109
    The Shooting of Delfin Lopez 110
    Labor Union Strikes and Strike Busting 113
    Delfin Lopez vs. Florencio Lasola 115
    Crispulo Atillo vs. Lauriano Sanchez 117
    Crispulo Atillo vs. Antonio Irogirog 119
    Death of Delfin Lopez 120
    Incarceration of Venancio Bacon 123
    Crispulo Atillo vs. Venancio Bacon 125
    The Split of Balintawak 129
    National Arnis Association of the Philippines 132
    Cebu Eskrima Association 133
    First National Open Arnis Championships 135
    First National Invitational Arnis Tournament 138
    Death of Venancio Bacon 140

CHAPTER 7: The Last Challenge 143
    DYLA Radio Interview 144
    The Contract 145
    The Main Event 146
    Aftermath and Rematch 150
    Missing Video Footage 152
    The Rematch 152
    The Cebu Coliseum 157
    Aftermath 159
    Ongoing Challenges and Failed Rematches 159

CHAPTER 8: A New Beginning 163

    Passing of an Icon 165

    Immigration to the United States 169

    Brothers of Balintawak 173

Angles of Attack 175

4 Methods of Defense 178

4 Methods of Attack 180

Mother Spar Drill with a Stick 182

Mother Spar Drill with Empty Hands 185

Disarm Techniques 187

Mother Spar with a Knife 195

Knife Disarms 197

Defense Against Hand Grabs 200

Family Tree 203

Atillo World Balintawak 204

Timeline of Events 214

Glossary of Terms 246

Memories 278

Bibliography and References 303

About the Authors 305

Forward

by Master Derrick Dalan
Student, Disciple, and Successor
Atillo Balintawak Eskrima - Original Saavedra Style

Grandmaster Atillo is the epitome of a true "Grandmaster." He lives and breathes the art of eskrima. Behind the skills, practicing, and teaching for over 63 years, I see the never-ending joy and passion of the art whenever he puts it on display.

He was already at the age of 70 when I first met him. I must say he is amazingly fast and precise with his technique, even up to this day. I remember calling my twin brother (Dennis Dalan) after the meeting, telling him that I just met "Master Yoda" of eskrima. I have never seen anybody move the way he does, young or old. As I train with him, I was even more mesmerized by his skills. He is always three to five steps ahead to counter ANY of your movement.

I am more accustomed to calling Grandmaster Atillo "Tatay" or "Tay" for short (meaning Father). I see him at least once a week to train. Since 2010, my brother and I have also assisted him whenever he conducts a seminar in the US. I have gotten to know him personally through the years. On our regular 3-day seminar trips, we would talk all night through the next morning about everything and anything, from WWII, famous eskrimador fight challenges, his family, his values, and of course, the art of eskrima and its history. I feel truly blessed to hear the story coming from the person who actually lived it.

Besides teaching through seminars, GM Atillo teaches mostly private lessons as he strongly believes in one-on-one teaching. Eskrima can be difficult to learn, especially in a group lesson setting, as he always says. He said learning it is like two people dancing. For a student to learn how to lead (in this case, to control someone's stick and hands), he or she must first dance with a person who knows how to lead. That is the reason why he makes a conscious effort to spar, touch sticks and hands, with every student who attends his seminar regardless of how small or big it is.

GM Atillo is a simple man who lives a simple life. He chose to live without lavish things. All he wants is to spread the art of Eskrima while providing for his family back home in the Philippines. He tells me, "People will not remember you by the material things possess, but they will remember you for the legacy you leave behind." He is the type of person who talks that talk and walks the walk. He is also very frank, a man who is not afraid to tell the truth regardless.

It is an honor, and a privilege, for my three sons (who may not know exactly how fortunate they are yet, because of their young age) and me, to know Grandmaster Atillo. He is a treasure. I am very thankful for his generosity, friendship, and his teaching.

Preface

I have been very fortunate to have had the opportunity to travel around the world and learn the indigenous fighting arts of many cultures. My desire to study and research the martial arts has created many opportunities and given me the proud privilege of becoming friends with some of the most talented and respected masters and instructors in the world.

The main focus of my research and training has been on the indigenous fighting arts of the Philippine Islands and the *pencak silat* systems of Indonesia and the Southern Philippines. Specifically, the methods of *Cebuano eskrima* developed and practiced on the island of Cebu in the central Visayan Islands of the Philippines

As I was preparing for another trip to Cebu City, Philippines, to continue my research, I was given the opportunity to visit and train with Crispulo "Ising" Atillo at his residence in Beaumont, California. Upon my arrival, Ising greeted me at the gated entrance to his home. I immediately recognized that he is an extremely friendly, humble, and sincere person who possesses a genuine desire to share his knowledge of Cebuano eskrima with anyone willing to learn. However, the moment I began my training, I realized he is not a typical instructor of Filipino martial arts. He is a professional *eskrimador* - a fighter. His realistic approach and extensive depth of knowledge reflect his decades of experience as one of the most illustrious and respected eskrimadors in the world.

As the youngest member of the original Balintawak Self Defense Club, founded in 1952 by Venancio "Anciong" Bacon, he is the last remaining link to the creation and early development of Balintawak eskrima. His lineage can be traced directly back to the legendary eskrimador, Lorenzo "Tatay Ensong" Saavedra himself and his celebrated nephew, Teodoro "Doring" Saavedra. It is from the hard-hitting style of Doring that Ising's unique style of Cebuano eskrima is derived.

Crispulo "Ising" Atillo has consolidated the knowledge he has gained over seven decades of Balintawak eskrima and combined it into a well-organized and systematic curriculum that he presents to his students in a logical and efficient learning progression. He also possesses a unique ability to transfer the knowledge he has gained from decades of real-world experiences in such a way the art can be immediately understood and applied by his students.

He is often referred to as the "Last of the Mohicans" and the "Last of the Gunfighters." As one of the few remaining legends of Balintawak eskrima, he embodies the true spirit of Filipino patriotism and his beloved art of Atillo Balintawak Eskrima - Original Saavedra Style. Like the Mohicans and gunfighters of the American Old West, he is the last of a very unique breed. Those remaining should be treasured, and their experience and knowledge captured and transferred to the next generation before they are inevitably gone.

Crispulo "Ising" Atillo has graciously shared his life and art with me for over a decade and has become a very close friend and mentor who I proudly call "Tatay." It is with great pride that I humbly accepted his request to collaborate with him and write this book to accurately document the history of Atillo Balintawak Eskrima - Original Saavedra Style. My only hope is that my efforts meet his expectations, and this book reflects the greatness befitting a true icon and living legend of the Filipino Martial Arts.

Glen Boodry
March 16, 2013

Acknowledgments

The authors would like to express their sincere appreciation to Magulang Na Guro Dan and Paula Inosanto of the Inosanto Academy of Martial Arts in Marina Del Rey, CA, for their support of Atillo Balintawak Eskrima and their tireless efforts to propagate the Filipino Martial Arts throughout the world. The authors would also like to thank Derrick Dalan and Dennis Dalan for their support and ongoing efforts to promote Atillo Balintawak Eskrima - Original Saavedra Style.

The authors would also like to thank Jason Gough for the incredible artwork and original cover design, Chris Callahan, for his efforts in proofreading and editing the original transcript, and Aliya Ferris for many of the photographs displayed in this book.

Crispulo "Ising" Atillo

I would like to thank Dr. Jesse Devera for his years of kindness and generosity, and Katherine Straus for her compassion and support. I would also like to thank Senator Juan Miguel "Migz" Zubiri, Mayor Bernard "Butch" Sepulveda, Ike Sepulveda, Macario Atillo, and Rudy Atillo for their assistance in the research for this book and their support of Atillo Balintawak Eskrima - Original Saavedra Style. I would also like to thank all of my Atillo Balintawak Eskrima students around the world whose loyalty and dedication to the art are greatly appreciated.

Glen Boodry

I would like to thank my family for their love and support. I would also like to thank the following friends and mentors for their influence and never-ending support throughout the years: Crispulo "Ising" Atillo, Guro Dan Inosanto, Paula Inosanto, Gokor Chivichyan, Burt Richardson, Larry Hartsell, Paul Vunak, Ramiro U. Estalilla, Jr., Thohsaphon Sitiwatjana, Bob Duggan, Cookie Vassiliou, Joel Clark, Tim Becherer, Michael Dubin, Doug Pechtel, Jon Saterstad, and Justin Buck. I'd like to especially thank Darrin Davis, Pete Italiano, and Tim Ferris for their never-ending friendship, encouragement, and support. I would also like to thank all of my students and the entire staff of the Inosanto Academy of Martial Arts.

Special thanks to:

Cecilia B. Cabantes
Curator, Museo Sugbo
Cebu Provincial Museum

Museo Sugbo
MJ Cuenco Avenue
Tejero, Cebu City, Philippines 6000

Hon. Hilario G. Davide, Jr.
Governor
Province of Cebu
Capitol, Cebu City, Philippines

May 10, 2011

Grandmaster Ising Atillo,

I am looking forward to the book you are writing with Glen Boodry entitled "Atillo Balintawak, "The Original Saavedra Style".

This book will be informative and give exposure to your Saavedra style-Balintawak system. Your system includes essential elements that should prove valuable to the Eskrima practitioner looking to expand his knowledge in Balintawak at the close quarter range.

This book will be an excellent addition to the library of any Eskrima practitioner

I have enjoyed the time we have been training together (since 2001), and wish you all the best with your new book.

Dan Inosanto
Founder/Head Instructor
The Inosanto Academy of Martial Arts
Marina Del Rey, CA USA

August 19, 2014

Grandmaster "Ising" Atillo,

It is with great pleasure that I congratulate Glen Boodry and yourself on the upcoming publication of your new book "Atillo Balintawak Eskrima: The Original Saavedra Style."

The foundation and techniques of Saavedra Balintawak you have introduced to our martial arts community has strengthened our skills and knowledge of Eskrima. It's a pleasure and great honor to train with one of the remaining original founding members of the Balintawak system.

Best wishes for continued good health and success. May this book help propagate the art of Atillo Balintawak to all interested practitioners.

Sincerely,

Ronald Goldstein
Student, Disciple, and Successor
Atillo Balintawak Eskrima - Original Saavedra Style

Prologue

The authors prefer to use the term *Cebuano eskrima* when generally describing the methods of *kali, eskrima,* and *arnis* practiced on the island of Cebu in the central Visayan Islands of the Philippines.

Although all methods of *kali, eskrima,* and *arnis* share some commonality, each method has a unique history that reflects the geographical region of the Philippines from where it originates and the distinct culture and personality of those who created it. These unique historical differences that define each method should be accurately documented and taught to all generations learning the art. If not, the history and physical characteristics that delineate one style from another will be dissolved into historical generalization and ultimately lead to the complete disappearance of the original art.

Furthermore, there are numerous styles, systems, and methods of Balintawak eskrima being practiced around the world today. The authors respect and appreciate all styles of Balintawak eskrima and encourage the reader to research and learn all methods available. Any style that can legitimately trace its lineage back to the Balintawak Self Defense Club founded in 1952, and the methods of eskrima practiced by the original members of the club, can rightfully use the name *Balintawak* and call itself Balintawak eskrima. Regardless of any modifications or contributions made to the original art practiced by the founding members.

In the words of Crispulo "Ising" Atillo - "We are all brothers."

Introduction

There are hundreds of styles, systems, and methods of the Filipino Martial Arts of *kali*, *eskrima,* and *arnis*. Each has its own history unique to the specific region of the Philippine Islands from where it originates and the individual masters who created it. Yet, all styles of *eskrima* share some historical and physical commonality that is central to all Filipino Martial Arts. However, some styles are inherently more practical and effective in modern combat and self-defense than others. This is particularly true of *Balintawak eskrima,* the style of eskrima originating from the Balintawak Self Defense Club founded in 1952 by Venancio "Anciong" Bacon.

Forged on the urban streets of Cebu City, Philippines, *Balintawak eskrima* is the result of Anciong's genius and the collective experience of several famed eskrimadors. Uniting together to preserve the combative nature of eskrima, many of these men were battle-hardened eskrimadors who served as Filipino guerillas during the war, streetfighters, gangsters, security guards, and members of the Cebu City Police. Several of these eskrimadors stand out in the early history of *Balintawak eskrima* - Vicente "Inting" Atillo, Delfin Lopez, Timoteo "Timor" Maranga, Jose "Joe" Villasin, and Teofilo "Pilo" Velez, to name a few. Then, of course, there is Crispulo "Ising" Atillo.

Of the remaining masters of *Balintawak eskrima*, Atillo is unique. Born on June 10, 1938, Crispulo Vestil "Ising" Atillo was literally born into a life of *eskrima* and raised by some of the most influential and renowned masters of the art. He is the son of renowned eskrimador, Vicente "Inting" Atillo, and the godson of the notorious Delfin Lopez, a feared eskrimador and Cebu gangster. As a child growing up in the Mambaling *barangay* of Cebu City, Philippines, he personally knew Lorenzo "Tatay Ensong" Saavedra, arguably the most influential eskrimador in the history of *Cebuano eskrima*, as well as Tatay Ensong's nephew, the legendary Teodoro "Doring" Saavedra. The latter widely considered the greatest fighter in history. Both Saavedra's were close family friends and neighbors who lived only footsteps away from Atillo's childhood home. In fact, before Atillo was old enough to earnestly learn eskrima, he would sit and watch Doring practice with his father, all-the-while mimicking Doring's movements with a stick. Atillo would say, "Doring, teach me!" Doring would laugh and offer the young

Atillo corrections and say, "Ising, you are a good boy. Someday you will be a good eskrimador!" Eventually, Doring became Atillo's beloved childhood mentor until his untimely death during World War II.

After the war, Delfin Lopez took the young Atillo under his wing. Lopez was a feared eskrimador and infamous gangster who could be ruthless and cruel toward his enemies, but loving and kind to those who respected and loved him. He had a particular fondness for the young Atillo. Not only did Lopez teach him *eskrima*, but also how to survive in the post-war Philippines plagued by corruption and violence. He groomed Atillo from the very beginning to be a fighter. A true eskrimador. As his uncle, Macario Atillo, once said during an interview shortly before his death: "Ising was a good boy, but he had a quick temper and was always getting into fights. He rarely picked fights, but he would never turn one down. He was just like Doring and copied his every move. He would walk the streets looking for trouble and opportunities to defend someone else just so he could fight."

It was also during this time in the early 1950s that Atillo learned the hard-hitting style of his childhood mentor Teodoro "Doring" Saavedra, directly from his father, Vicente "Inting" Atillo. Inting was a life-long friend of Doring's and served alongside Doring and Delfin Lopez during World War II as a member of the famed guerilla forces under Lt. Col. James Cushing. He was also a childhood friend and training partner of the renowned Venancio "Anciong" Bacon, the godson of Atillo's grandfather. Anciong lived nearby and was a frequent visitor at the Atillo residence. Following Doring's death, he was widely considered the best eskrimador in Cebu and the second greatest fighter in the history of *Cebuano eskrima*. As a young 14-year-old student of eskrima, Atillo was present during many of the discussions leading up to Anciong's decision to leave the famed Doce Pares Club and was witness to the founding of the celebrated Balintawak Self Defense Club in 1952. As the youngest member of the club, he would participate in weekly training sessions with Anciong and his father in a vacant lot adjacent to his home in Mambaling. He would often be asked by Anciong to participate in demonstrations at local fiestas alongside older and more experienced eskrimadors. Many of whom were famous eskrimadors in their own right.

For the next several decades, Atillo witnessed nearly every major historical event in the history of *Balintawak eskrima,* including many of the famed *juego todo* matches that occurred between rival eskrimadors. He also participated in several formal matches to include fights with renowned Doce

Pares Club eskrimadors, Lauriano Sanchez and Antonio Irogirog. In due course, Atillo's keen sense of honor and loyalty led him to an unexpected and controversial *juego todo* match with Anciong himself and a contentious fight with the famed Ciriaco "Cacoy" Cañete of the Doce Pares Club. His fight with Cacoy is still talked about to this day. It is recognized as the last formally sectioned challenge-match between eskrimadors.

This is not another book of step-by-step techniques or one that simply repeats the same previously published history of *Balintawak eskrima* that has been told countless times before. Instead, it is the story of Crispulo "Ising" Atillo and the little-known history of the Atillo Balintawak Eskrima-Original Saavedra Style as told through the eyes of one of the most celebrated and respected icons and masters of the Filipino Martial Arts.

CHAPTER 1

Lorenzo Saavedra

Lorenzo "Tatay Ensong" Saavedra was born in the city of Carcar south of Cebu City in 1852 and is perhaps the most prominent and influential eskrima master in the history of Cebuano eskrima. His lineage and teachings directly influenced the history and development of many of the present-day styles of eskrima practiced throughout the world today. It is not known where, or from whom, *Tatay* Ensong initially learned eskrima. However, many believe that Tatay Ensong's original method of eskrima was taught to him by elder members of the Saavedra family. It is also likely that he encountered and learned from other eskrimadors living on the island of Cebu and traveling throughout the region during the latter part of the 19th Century. Even at an early age, Tatay Ensong was considered a very talented and accomplished eskrimador. Especially with the *solo olisi* and *punta y daga* method of eskrima, which were described by his later students as *corto linear*, *de cadena,* and *de cuerdas*.

The legendary Lorenzo "Tatay Ensong" Saavedra (center), Cebu City, Philippines, c. 1932.

National Uprising and Call for Independence

Throughout the Spanish Colonial Era of the Philippines,[1] the possibility of

---

[1] The Spanish Colonial Era of the Philippines began with the arrival of Ferdinand Magellan in 1521 and official colonization beginning with the arrival of Miguel López de Legazpi in 1565.

indigenous uprisings and external invasions by the Dutch, Chinese, Japanese, and British were an ongoing concern for the Spanish governing authorities. On January 20, 1872, approximately 200 colonial soldiers and general laborers stationed at the Spanish arsenal at Fort San Felipe south of Manila in Cavite el Viejo revolted against the Spanish government and seized control of the fort. Led by *mestizo* Sergeant Francisco La Madrid, the mutineers hoped to inspire a national uprising and call for Philippine nationalism.

The cause of the mutiny is believed to be the result of an order issued by Governor-General Rafael de Izquierdo y Gutiérrez[2] to subject soldiers of the Engineering and Artillery Corps to a personal tax. The order issued by Gutiérrez required the soldiers to pay a monetary sum to the Spanish government and perform *polo y servico*[3]. The mutiny sparked when the soldiers and laborers realized the tax and *falla*[4] had been deducted from their pay.

The next day, General Felipe Ginovés surrounded Fort San Felipe with an entire regiment of Spanish soldiers. They besieged the fort until Madrid, and the mutineers surrendered. Ginovés then ordered the soldiers to fire upon the mutineers as they exited the confines of the fort to surrender. Those not immediately killed were arrested and imprisoned or deported to one of the many Spanish penal colonies throughout the region. The Spanish authorities also arrested and executed anyone believed to have directly supported the mutineers.

On January 27, 1872, Governor-General Rafael Izquierdo approved the death sentences of forty-one of the mutineers. On February 6, 1872, eleven more were sentenced to death, although these sentences were later reduced to

---

The period ended with the signing of The Treaty of Paris on December 10, 1898, at which time Spain relinquished all claim of sovereignty over the Philippines to the United States

[2] Rafael de Izquierdo y Gutiérrez was a Spanish Military Officer, Political Leader and Statesman who became Governor-General of the Philippines from April 4, 1871, to January 8, 1873. He was famous for his use of "Iron Fist" type of government and was the Governor-General during the 1872 Cavite mutiny which led to the execution of 41 of the mutineers, including the Gomburza martyrs.

[3] A system of forced labor which evolved within the framework of the encomienda system, introduced into the Philippines by the Conquistadores and Catholic priests who accompanied them.

[4] A fine paid to the Spanish government to exempt oneself from forced labor.

life imprisonment. Others were exiled to Guam and the Mariana Islands. The most significant group settled in Europe and established a colony of Filipino expatriates in Madrid and Barcelona, Spain. These proud Filipino patriots were able to create small associations and print publications that encouraged the Philippine Revolution.

Afterward, Spanish authorities criminally charged three Filipino priests, Mariano Gómez, José Burgos, and Jacinto Zamora of *subversion* arising from the Cavite Mutiny. The Spanish authorities summarily sentenced the priests to death and publicly executed them by *garrote*[5] in Bagumbayan[6] on February 17, 1872. Their brutal and callous execution left a profound impact on many Filipinos, including José Rizal[7], the national hero of the Philippines. It inspired a national consciousness that became the impetus for many Filipinos to unite and condemn the abuse of the colonial authorities.

The National hero of the Philippines, José Rizal fencing with Juan Luna and Mariano Ponce in Paris, France in 1883.

Fearing that a colony-wide rebellion was being planned to overthrow the Spanish government in the Philippines, authorities cracked down on the rapidly increasing nationalist movement. They began to arrest, imprison, exile, and execute anyone who spoke out publicly against the colonial government or supported anti-government activities. Additionally,

---

[5] A cruel device used by Spanish authorities to execute prisoners sentenced to capital punishment. The device consisted of a seat and back brace that restrained the victim while a metal band was tightened around their neck with a crank until the person died from strangulation or a broken neck.

[6] The township of Bagumbayan is now Rizal Park, also known as Luneta Park, an urban park located on Roxas Boulevard in the capital city of Manila, Philippines.

[7] José Protacio Rizal Mercado y Alonso Realonda (June 19, 1861 – December 30, 1896) was a Filipino nationalist, writer and revolutionary who is widely considered the greatest national hero of the Philippines.

the crackdown brought about efforts by the Spanish to disarm Filipinos by issuing a decree banning the possession of any weapons. Until then, many Filipinos owned indigenous bladed weapons inherited, fabricated by a local *panday*, or acquired through theft and trade. Additionally, many were in possession of different revolvers and rifles stolen or obtained through the black market. These weapons were typically kept hidden by the owner to prevent them from being stolen by thieves or confiscated by the local authorities. Anyone caught in possession of a weapon, other than a bladed tool strictly used for agricultural purposes, would be accused of being a rebel, incarcerated, and most likely executed.

The incident, later known as the Cavite Mutiny of 1872 and the subsequent public execution of the three Filipino priests, is generally thought to be the beginning of the Philippine Revolution.

Rebel Accusation and Incarceration

That same year, the Spanish authorities accused Tatay Ensong of being a rebel and charged him with *insurrection*. Like many young Filipinos at the time, he supported the growing nationalist movement and actively spoke out against the colonial government. He was arrested by the *Guardia Civil* and incarcerated in the Cebu Provincial Jail[8] by the *Sala de lo Criminal* for an indefinite number of years. The jail was constructed in 1871 and was initially

The Cebu Provincial Jail during its final construction just prior to the incarceration of Lorenzo Saavedra in 1872.

---

[8] The Cebu Provincial Jail, also known as the Cárcel de Cebú and the Cebu Provincial Detention and Rehabilitation Center (CPDRC). Designed by Domingo de Escondrillas in 1869 to be the main prison of the Visayas District, it was constructed in 1871. It is now the site of the Museo Sugbo, the Cebu Provincial Museum and is a popular tourist destination in Cebu City.

designed to be the *Cárcel del Distrito* and imprison criminals throughout the Visayan region. The penitentiary was later used to incarcerate members of the *Katipunan*[9] during the Philippine Revolution and Filipino guerillas during the Japanese occupation of the Philippines during WWII.

As a child, I often heard stories of Tatay Ensong's support of the nationalist movement and his rebel activities during his youth. Many believed he was a member of the Katipunan; however, that would not have been possible. Andres Bonifacio officially established the Katipunan in Tondo, Manila, on July 7, 1892, ten years after Tatay Ensong's incarceration in the Cebu Provincial Jail. Nonetheless, according to those who personally knew Tatay Ensong at the time, he was an active supporter of the early nationalist movement. He was actively involved in rebel activities against the Spanish authorities governing the Philippines islands.

The French School of Fencing

While incarcerated in the Cebu Provincial Jail, Tatay Ensong befriended a French prisoner who was a practitioner of the French method of traditional European sword fighting, or *escrime*. Tatay Ensong found the techniques of the Frenchman very convincing and difficult to counter using the traditional eskrima techniques he had previously learned. Frustrated, he was determined to learn the Frenchman's techniques so he could

Bilibid Prison, Muntinlupa, Philippines, 1925. (Photo Courtesy the National Oceanic and Atmospheric Administration, NOAA Photo Library, and family of Captain George L. Anderson).

counter them and improve his own method of eskrima. Tatay Ensong became the Frenchman's student and began learning the sword and dagger techniques

---

[9] The Katipunan was a Philippine revolutionary society founded by Filipino patriots Andres Bonifacio, Teodoro Plata and Ladislao Diwa on July 7, 1892, in Manila, Philippines. Its primary goal was to achieve Philippine independence from the Spanish through revolution.

Cebu Provincial Jail inmates hearing Mass, October 10, 1947. (Photo courtesy of the Museo Sugbo, Cebu City, Philippines).

of the French School of Fencing from his fellow prisoner. The French School of Fencing, or *Ecole Française d'Escrime*, is a regionally indigenous sword and dagger art heavily influenced by the 16th-century Italian styles of rapier fencing. Also known as the *Académie des Maistres en faits d'armes de' Académie du Roy*, The French School of Fencing was formally established in December 1567 by King Charles IX of France. This method of fencing is believed to be the original method practiced by the Paladins, or *Doce Pares de Francia*, the Twelve Peers of France, the foremost warriors of King Charlemagne's court. Charlemagne, or Charles the Great, was the King of the Franks from 768 and Emperor of the Romans from 800 until he died in 814. In 1573, 16th-century French fencing master, Henry de Sainct-Didier authored and published what is believed to be the first French fencing manual ever written. During his 25-year career in the French army, de Sainct-Didier participated in the Great Wars of Italy[10] and learned the Italian method of swordplay. In his manual, de Sainct-Didier states he "lived his whole life learning to fight with the single sword," His reason for writing the book was

---

[10] The Great Wars of Italy, or Italian Wars, were a long series of wars fought between 1494 and 1559 in Italy during the Renaissance period.

to "further serve his king." Fencing in France was then developed into a sport during the 17th century, with the codification of rules and terminology and a system of teaching, by masters such as Jean Baptiste la Perche du Coudray, Charles Besnard, Philibert de la Touché, and Jean Labat of Toulouse. The French method of fencing employed a smaller weapon that facilitated fast and elegant thrusting motions rather than an actual dueling sword. French fencers blunted or foiled the point of the sword by wrapping foil around the blade or fastening a knob at the tip of the blade to make the practice safer. These practices ultimately lead to the invention of the fencing foil in France in the middle of the century. By the 18th century, the French school of fencing had become the western European standard for fencing. The emergence of classical sport fencing in the 19th century was a direct continuation of the French tradition.

Development and Refinement During Incarceration

Unfortunately, the actual name of the Frenchman is unknown, and it is not known why the Spanish authorities incarcerated him in the Cebu Provincial Jail. It is also unknown what became of him after Tatay Ensong's release from prison. The existence of French citizens in the Philippines was not uncommon, and the incarceration of a Frenchman in a Spanish colonial prison would have been very possible. Especially for

The Cárcel de Cebú, then known as the Cebu Provincial Jail, c. 1915 during the incarceration of Lorenzo "Tatay Ensong" Saavedra. (Photo courtesy of the Museo Sugbo, Cebu City, Philippines).

insurrection or supporting the growing nationalist movement. Efforts to reform and gain independence from Spanish rule coincided with the French

Inmates of the Cebu Provincial Jail (in black and white stripes, the standard US colonial era prison attire) digging trenches, c. 1902. (Photo courtesy of the Museo Sugbo, Cebu City, Philippines).

Revolution [11] and the social and political upheaval in France and French colonies throughout the world. Including the French trading ports that existed throughout the East Indies and Philippine Islands. Additionally, France maintained a close relationship with the Spanish colonial government. Despite Spanish restrictions banning foreign trade, French traders were already in the Philippines long before the islands were opened to international trade.

When not under the watchful eye of the Spanish prison guards, the two *escrimeurs* would practice at night in secluded areas of the prison where they could segregate themselves from the other prisoners. Since bladed weapons were not readily available for practice, Tatay Ensong and the Frenchman practiced using available materials to create makeshift weapons

---

[11] The French Revolution (May 5, 1789 - November 9, 1799) was a period of great social and political upheaval in France and French colonies around the world. It profoundly altered the course of modern history and triggering the global decline of absolute monarchies while replacing them with republics and liberal democracies.

for training. This, in all probability, forced them to modify and adjust their techniques to fit the assortment of improvised implements they made. In addition, the harsh and challenging conditions of prison life, coupled with a constant threat of inmate violence and prison rioting, likely forced Tatay Ensong and the Frenchman to use the improvised weapons for self-protection and further adapt their techniques to close combat as a means of survival.

Dueling under the reign of King Henry IV of France using the sword and dagger method.

Throughout his years of confinement, Tatay Ensong continued training and eventually mastered the techniques taught to him by the Frenchman. He adopted many of the sword and dagger techniques of the French School of Fencing into his method of eskrima. Today, it is believed that the techniques taught to Tatay Ensong by the Frenchman encompass over sixty percent of his unique style of *Cebuano eskrima*.

It is here, in the Cebu Provincial Jail, where Lorenzo "Tatay Ensong" Saavedra was to spend the majority of his life.

CHAPTER 2
Labangon

Shortly after his release from the Cebu Provincial Jail, Lorenzo "Tatay Ensong" Saavedra returned to live with his family in the San Nicolas District of Cebu City near my home in Mambaling. He began teaching his innovative method of *Cebuano eskrima* to his nephews, Federico "Pedring" Saavedra and Teodoro "Doring" Saavedra, who had been born roughly ten years earlier on October 24, 1911. As interest in learning the art grew, Tatay Ensong began teaching several other dedicated friends and interested youth in the surrounding districts of San Nicolas, Labangon, and Mambaling.

San Nicolas, Cebu, Philippines, c. 1930.

The exact date of Tatay Ensong's release from prison is unknown, and prisoner records maintained at the Cebu Provincial Jail unfortunately no longer exist. According to Jojo R. Bersales, Ph.D., who served as a museum consultant for the Province of Cebu between 2008 and 2012, all records previously held at the prison were sadly lost or destroyed. Most were lost during the many transitions the prison went through; however, some remaining documents indicate many prisoners were released following the signing of Proclamation 483 by United States President Theodore Roosevelt on July 4, 1902. The Proclamation granted a full pardon and amnesty to all persons in the Philippines who participated in or supported insurrection activities against Spain and the United States between August 1896 until the cession of the Philippines by the United States in 1898. Some historians have suggested that prisoners convicted and incarcerated by Spanish authorities

South Road, Cebu, Philippines, c. 1920.

earlier than 1896 may have also been pardoned and released shortly after implementing the proclamation in 1902. However, it is most likely that Tatay Ensong was released following the signing of the Philippine Autonomy Act of 1916, or the Jones Act, shortly before the completion of Francis B. Harrison's term as the American governor-general of the Philippines in 1921. The Jones Act was enacted by the United States Congress on August 29, 1916, and formally declared the intention of the United States to grant independence to the Philippines as soon as a stable government could be established. Harrison moved quickly to replace Americans with Filipinos in the civil service. By the end of his term in 1921, Filipinos had taken charge of all internal affairs in the Philippines. This account matches the limited historical records available regarding the Cebu Provincial Jail and accurately reflects the oral history conveyed by Tatay Ensong and his early students at the time.

Labangon Fencing Club

As public interest in eskrima increased and people began openly practicing and sharing their knowledge of the art, Tatay Ensong recognized the benefits of bringing *eskrimadors* together into a single organized group to practice and exchange ideas as well as pass the art onto the next generation of interested youth. Tatay Ensong began scheduling organized practice sessions and formally established the Labangon Fencing Club on August 14, 1920.

Recognized as the first formal eskrima club in the Philippines, the Labangon Fencing Club was also informally referred to as the Lantay Club for the bamboo bench or *lantay* where students would sit and wait their turn to practice. To some extent, it was an informal gathering of friends and family who shared a common interest in practicing the art. In addition to Tatay Ensong and his nephews, Federico and Doring Saavedra, other original club members included Rostico Covarrubias as well as 19-year-old Eulogio "Yoling" Cañete and his 15-year-old brother Filemon "Momoy" Cañete.

There were, in fact, other groups of eskrimadors practicing eskrima throughout the island of Cebu at the time the Labangon Fencing Club was founded. In addition to Tatay Ensong, Estanislao "Islao" Romo of Pasil, Cebu City, founded an eskrima club in the neighboring barangay of Tisa in 1921, the Ilustrísimo family were sharing their art in Bagong, Bantayan, in the Northern part of Cebu.

Islao Romo, Tisa, Cebu City, Philippines, c. 1927.

It is often said that Venancio "Anciong" Bacon and Ciriaco "Cacoy" Cañete[12] were both members of the Labangon Fencing Club when the club was founded in 1920. That is not entirely correct. Anciong was born on

---

[12] Ciriaco "Cacoy" Cañete (August 8, 1919 - February 5, 2016) was a member of the Doce Pares Club and founded Eskrido, his own personal system of eskrima in 1951.

October 15, 1912, and was seven years old when it was founded. Anciong was a close childhood friend and training partner of Doring and was taught by Tatay Ensong alongside Doring. However, Anciong and Doring were very young children. They were most likely taught informally by Tatay Ensong until they were old enough to practice in earnest. My dad, Vicente "Inting" Atillo, was also born the same year as Anciong on March 15, 1912, in Mambaling, and didn't begin practicing with Doring until years later. Cacoy was born August 8, 1919, and would have been an infant when Tatay Ensong founded the Labangon Fencing Club.

Translation and Interpretation

There have been many explanations regarding why Tatay Ensong chose the term *fencing* rather than *eskrima* when naming the club, the Labangon Fencing Club. Many believe that Tatay Ensong chose the term *fencing* because it was the word used by the American regime to describe what was being practiced by the native Filipinos since it closely resembled the European sport of fencing. This is not entirely correct. Eskrima has always been called *fencing* in the English language and still is to this day. The term *fencing* is just the English translation of the Filipino term *eskrima*, just as *fencing* is the English translation of the Spanish term *esgrima*, French term *escrime*, Portuguese term *esgrima*, and Italian term *scherma*. The Americans encouraged Filipinos to speak their native language and learn the English language during the American occupation.

Styles and Methods

The art practiced by the members of the Labangon Fencing Club was referred to as *eskrima* and not categorized into different styles like today. General descriptions were used to describe an individual's approach to eskrima, such as *corto linear*, *riterada*, and *larga mano*, but eskrima was just *eskrima*. Members of the club who had previous training would share what they had learned from others and share any personal contributions or developments they had made. Eskrimadors had been sharing information and developing their own methods before establishing the Labangon Fencing Club. Still, the club is generally believed to be the first time eskrimadors and interested youth were invited to come together and practice the art as an organized group.

Additionally, the majority of the eskrima practiced at the time still employed blade-oriented techniques adopted from Spanish sword fighting

and indigenous blade fighting methods. Although a bladed weapon was still practiced within the art, it was most often taught with wooden or rattan sticks and replica weapons. Training with actual blades was kept hidden due to the potential of being considered a rebel or a common criminal. Before the liberation of the Philippines from the Spanish by the United States, unless a native Filipino was a member of the Spanish Colonial Army or Civil Guard[13], possession of a bladed weapon, other than those strictly used for agricultural purposes, could have led to incarceration and potential death.

A young 18-year-old Floro Villabrille. Sr. (1912 - 1992) demonstrating early espada y daga techniques, c. 1930.

Because of that, eskrimadors were very untrusting of the prevailing government authorities and extraordinarily guarded and careful when practicing or teaching the art and use of bladed weapons. Instead, many of the blade-oriented techniques were modified and adapted for use with sticks of various lengths, sizes, and configurations that simulated bladed weapons. Tatay Ensong had been incarcerated in the Cebu Provincial Jail most of his life for insurrection and was forced to use wooden implements to simulate bladed weapons during practice. Out of pure necessity, his tactics and techniques evolved from being blade oriented to those more applicable to an impact weapon. Even the *punta y daga*[14] techniques taught by Tatay Ensong following his release from prison

---

[13] The Civil Guard, or Guardia Civil, was a component of the Spanish Army organized under the Spanish colonial government in the Philippines to enforce civil law, and impose penalties for infringements of law and local ordinances. It was disbanded after the Spanish - American War and replaced by the Philippine Constabulary.

[14] Punta y daga, meaning *point and dagger*, is an alternate term used to describe the espada y daga, or *sword and dagger*, techniques of eskrima adopted from the Spanish. Particularly

employed more thrusting and striking motions rather than slashing and cutting techniques because of a lack of a cutting edge when using a wooden or rattan implement.

Quarreling and Disagreements

In the late 1920s, the Labangon Fencing Club struggled due to supposed political infighting and quarreling between its members. Maintaining a cooperative group focused on the pure enjoyment of practicing eskrima proved to be a challenging task due to constant bickering and infighting amongst its members. A formal board of directors was established to institute a hierarchy and establish control over practice sessions. Eventually, a schedule of practice times and locations was created to separate the quarreling groups and prevent conflicts between members of the club. Regrettably, this separation was contradictory to Tatay Ensong's intent and eventually began to lead to a break down in the club. Personal biases began to grow regarding who were the better eskrimadors and whose method of eskrima was superior. This was subsequently plunging the club into a constant state of discouragement and conflict. It was said that while one student was practicing and demonstrating, the other students watching would make fun and criticize him. When that student finished, someone else would suffer the same harassment. This constant teasing and ridicule eventually lead to the club being offhandedly referred to as the *iring ug iro* club or club of "cats and dogs."

Formal Disbandment

Tatay Ensong and the participating members of the club were financially very poor. They relied heavily on monetary donations to continue operating. In one unfortunate instance, 500 pesos went missing from the Mayor of Cebu City to support the club. According to Eulogio "Yoling" Cañete, who served as the club's Secretary, the loss of the money caused a stir within the club, and members began accusing one another of theft. Cañete further said that Tatay Ensong refused to discuss the issue, which failed to resolve the situation. Accusations and counter-accusations worsened and continued until angered members would no longer participate in practice.

---

when referring to the use of a wooden or rattan stick for thrusting rather than a sword.

The issue was never resolved, and a vote was held by the remaining board members. They elected to officially disband the Labangon Fencing Club on August 14, 1930, exactly ten years from the date it was founded.

CHAPTER 3
The Doce Pares Club

Even though the Labangon Fencing Club had been disbanded in 1930, Lorenzo "Tatay Ensong" Saavedra continued teaching his devoted students who were now older and more mature eskrimadors. His nephew, Teodoro "Doring" Saavedra, was now 19 years old and had established himself as a formidable fighter and seasoned eskrimador. Venancio "Anciong" Bacon was a year younger and had been training with Tatay Ensong alongside Doring for several years. Anciong had also become a very talented and tough eskrimador and was known as the second-best fighter within the group next to Doring.

Teodoro "Doring" Saavedra, 1932.

A short time later, Doring approached Tatay Ensong together with Eulogio "Yoling" Cañete and Filemon "Momoy" Cañete and proposed forming a new club. Although there were adult members of the former Labangon Fencing Club, it mostly consisted of teenagers and children. Instead, Doring suggested that the new club be primarily composed of adults interested in serious combative training rather than an informal gathering of children. Many of Tatay Ensong's students were attracted to the idea, and serious discussions to create a new club began toward the end of 1931.

Over the next few months, several meetings were held to discuss the formation of the club. Tatay Ensong and the original organizers wanted to give the new club a more dramatic and recognizable name. One that did not reflect a specific location, as was the case with the former Labangon Fencing Club. During one such meeting on January 11, 1932, the group acknowledged twelve principal organizers and soon-to-be members were in attendance. After some discussion and debate over several potential names, it was suggested by Tatay Ensong that the new club be called the *Doce Pares Club*.

Tatay Ensong chose the name to honor the Frenchman he befriended while incarcerated in the Cebu Provincial Jail. Additionally, the name represented the *Doce Pares de Francia*, or Twelve Peers of France, the twelve legendary swordsmen of King Charlemagne of France. The Twelve Peers were the elite paladins and foremost warriors of Charlemagne's imperial army and his closest advisors. The number of peers is also thought to parallel the twelve apostles of Jesus Christ. Additionally, the story of the *Doce Pares de Francia* was a well-known *corrido*[15] throughout the Philippine Islands. The story portrayed an extended account of the legendary exploits of King Charlemagne and his twelve warrior swordsmen.

The meaning of *doce pares* has been a point of debate over the years, and it has evolved into meaning many different things. Much of which is not historically accurate. Truthfully, a long-standing misunderstanding has existed for decades regarding the meaning of the term. This was most likely caused by a mistranslation or misinterpretation by those passing down the oral history of the Doce Pares Club to the later generations.

Eulogio "Yoling" Cañete, c. 1932.

The term *doce pares,* refers to the twelve *peers* of France and not twelve *pairs,* as many people have suggested. Although the term can be loosely interpreted to mean *twelve pairs*, it does not refer to two groups of twelve or twenty-four persons at all. It refers to the equality amongst the twelve paladins of King Charlemagne as noblemen. Many eskrimadors suggested that Tatay Ensong and the founding members of the Doce Pares Club intentionally chose the name to represent two pairs of twelve. It is further suggested they achieved their goal when the club's membership increased to twenty-four members. Some have even claimed that the name was chosen to reflect the twelve basic strikes that are common in most styles of eskrima. The correlation between the name and the increase in membership to twenty-four members was an afterthought and coincidence.

---

[15] A *corrido* or *korido* is a metrical story about the life and adventures of a person usually sung in the accompaniment of a guitar.

Membership Growth and Controversy

As soon as the Doce Pares Club was founded, membership in the club increased. At that time, the members of the club held a formal election of officers on January 21, 1932. According to my dad, Vicente "Inting" Atillo and Fortunato Peñalosa, who acted as the Secretary of the club from the time it was established in 1932 until it was temporary disbandment due to the Japanese invasion of Cebu in 1941, the following were original members of the club: Eulogio "Yoling" Cañete (President), Teodoro "Doring" Saavedra (Vice President), Federico "Pedring" Saavedra (Director), Florentino Cañete (Director), Magdaleno Cabasan (Director), Strong Tupas, (Director), Juanito Lauron (Director), Filemon "Momoy" Cañete (Sergeant at Arms), Venancio "Anciong" Bacon (Sergeant at Arms), Rodolfo Quijano (Sergeant at Arms), Fortunato Peñalosa (Secretary), Marcelo Verano (Treasurer) and Deogracias Nadela (Auditor). Other members serving on the Advisory Board included Dr. Anastacio Deparine, Atty. Margarito Revilles, and Atty. Cecilio de la Victoria. My dad joined a short time later, along with Jesús Cui, Pio Deparine (Historian), Victorino Dilao, José Garcia, Rosalio Gonzales, Claudio Kalinawan, Francisco Roncesvalles, Basilio Tumulak, Pastor Villagracia, and Felipe Villaro.

The history of King Charlemagne and the Twelve Peers of France by José Vazquez, 1837.

Significantly older than the others at 80 years of age, Tatay Ensong served as the club's chief instructor; however, he mostly acted as a mentor to the younger instructors within the club. Doring served as the club's chief instructor when the elder Saavedra was not there. Doring's fighting style was hard-hitting and aggressive. Known as one of the best fighters on the island of Cebu, Doring's fighting exploits were already becoming well known throughout the Philippines.

Elected officers of the Doce Pares Club in 1932. 1st row, 4th from left, Eulogio "Yoling Cañete (President), 5th from left, Lorenzo "Tatay Ensong" Saavedra (Head Instructor), 7th from left, Filemon "Momoy" Cañete (Sergeant at Arms). 2nd row 2nd from left, Teodoro "Doring" Saavedra (Vice President).

Although Peñalosa's records undoubtedly documented many founding members of the Doce Pares Club, the original membership list has always been a point of controversy. Including the name of the first president of the Doce Pares Club, who many believe to be Rev. Rendal. Additionally, some unaffiliated students considered themselves members of the Doce Pares Club through their association with an instructor or another genuine member. Furthermore, there were offshoot clubs throughout the Philippines that began illegitimately calling themselves the Doce Pares Club. This was caused by the increasing popularity of eskrima and the growing reputations of many legitimate club members like Teodoro "Doring" Saavedra.

Teodoro Saavedra vs. Pablo Alicante

Shortly after the Doce Pares Club was founded, an officially sanctioned match between Estanislao "Islao" Romo and Pablo "Ambong" Alicante was arranged. The winner of the match would be declared the eskrima champion of Cebu. The fight was sanctioned by the mayor of Argao, a coastal town south of Cebu City, and was scheduled to occur in September of 1933 during a town fiesta. A challenge match or *bahad* was a popular way for rival eskrimadors to challenge one another to determine who was the better fighter and resolve personal rivalries and conflict. Challenge matches required the approval of local officials. Also, each fighter was required to sign a contract agreeing to the rules of the match. The signing also released each fighter from any legal responsibility for injuries or death sustained during the match.

Considered one of the best fighters in the region, Romo was not an official member of the Doce Pares Club. However, he was close friends with several club members, including Tatay Ensong and Doring. Romo and Tatay Ensong mutually respected one another and held each other in very high esteem. Residing in the Pasil District of Cebu City, he was a well-respected and feared eskrimador who had proven his ability as a fighter. Alicante was a fierce eskrimador who openly challenged anyone who thought they could beat him in a match. Alicante alleged that everyone was afraid of him and often bragged that nobody could defeat him. Alicante was also known to use *anting-anting*[16] during his fights to protect him against physical injury and defeat.

Romo initially accepted Alicante's challenge but canceled the night before stating he had not discussed the match with his wife, and she feared for his safety. Alicante mocked Romo's decision. He began ridiculing him publicly, claiming he was using his wife as an excuse because he was afraid to fight. Angered by Alicante's arrogance, Doring volunteered to replace Romo and accepted Alicante's challenge.

On the day of the event, a large crowd of spectators gathered to witness the match between the two well-known and celebrated eskrimadors. My dad was there, as was my uncle Macario Atillo and Tatay Ensong. The latter acted as the cornerman for his nephew and prized pupil.

As the referee signaled for the first round to begin, Doring froze in place and suddenly couldn't move. Fearing that Doring was under the influence of Alicante's anting-anting and was vulnerable to Alicante's attack, Tatay Ensong shouted at Doring and yelled out, "*Bantay*![17]" Doring instantly

---

[16] A Filipino term that refers to charms, talismans and amulets believed to possess magical powers commonly used for good luck.

[17] Bantay is a Tagalog term for "look out" or "to be on guard."

snapped out of it just as Alicante was swinging his stick toward his head. Narrowly avoiding the impact, Doring defended the strike with his stick, which nearly broke in two from Alicante's powerful blow. The two eskrimadors continued fighting until the end of the round, which was clearly won by Alicante.

With increased confidence, the second round began with Doring on the offensive, striking Alicante repeatedly in the head and body. Astonishingly, with the help of his anting-anting, Alicante removed a handkerchief from his pocket and wiped away the bruises caused by Doring's blows. Seeing the bruises disappear as if his strikes had no effect on Alicante played a psychological toll on Doring. He returned to his corner at the end of the second round feeling frustrated and belittled. Until now, the match was very close, and Doring needed a decisive round to win. Recognizing that Alicante was holding something in his mouth, Tatay Ensong whispered into Doring's ear and instructed him to strike the object out of Alicante's mouth.

Street-side amulets commonly used as anting-anting throughout the Philippines.

As the final round began, the two eskrimadors clashed in a fierce exchange of strikes. Suddenly, Doring struck Alicante's mouth, causing him to spit out the object he was hiding. It was a small crucifix of Jesus Christ that he had been using as his anting-anting. Suddenly, blood began spewing from Alicante's mouth. Doring then noticed the bruises that Alicante had wiped away with his handkerchief were now clearly visible all over his head and body. Alicante was noticeably hurt and no longer being protected by his anting-anting. Doring quickly rushed forward with an aggressive attack and began repeatedly striking Alicante. No longer under the protection of his anting-anting, Alicante raised his hands in defeat and surrendered. The referee stepped between the eskrimadors to protect Alicante and declared Doring the champion of Cebu. After the match was over, Alicante accepted that he had been defeated and shouted, "Doring! You are the first ever to beat me!"

The fight with Alicante became known as the most memorable match of Doring's career. It established him as one of the most feared eskrimadors in the Philippines. In addition, Doring's defeat of Alicante solidified the reputation of the Doce Pares Club and popularized the name throughout the Philippines.

Macario Atillo, 2013. Witness to many fights including the match between Teodoro "Doring" Saavedra and Pablo "Ambong" Alicante. Sadly, Macario passed away weeks after being interviewed by the author.

Teodoro Saavedra vs. Pedrito Romo

In 1937, a few years after Doring's fight with Alicante, Tatay Ensong and Doring again approached Estanislao "Islao" Romo. They invited him to join the Doce Pares Club. Romo created his own club years earlier in the 1920s and was apprehensive about partnering with Tatay Ensong even though they were friends. Romo stated he would join the Doce Pares Club under one condition. Doring and his son Pedrito "Pedring" Romo would have to fight to see who was the better eskrimador. Romo advised that if Pedring lost to Doring, he would join the Doce Pares Club. Doring and Pedring agreed and squared off in a friendly contest to determine who was the better eskrimador.

Just as the two eskrimadors prepared to begin, Pedring thrust his stick forward and struck a surprised Doring in the forehead. Tatay Ensong immediately yelled at Doring and instructed him to fight. Doring rushed forward and began striking Pedring, who was quickly overwhelmed by Doring's attack. Pedring continued backing away in an attempt to defend himself until he eventually tripped backward and fell over several *sabong*[18] stacked behind him. Recognizing the match was over, the two young eskrimadors acknowledged Doring had won and was the better fighter. Although Doring had decisively defeated his son, Romo stubbornly maintained his position and elected not to join the Doce Pares Club.

Cockfighting in the Philippines, c. 1940s.

Teodoro Saavedra vs. Roman Ladaño

The following year in 1938, Doring fought another eskrimador from Negros named Roman "Oman" Ladaño. The match took place near my home in Basak in front of F. Llamas Street. The fight lasted only a few seconds and resulted in Oman being knocked out by Doring. Once Oman regained consciousness, he angrily yelled, "Doring! We will fight again!" Confident in his ability as an eskrimador, Doring responded and said, "Anytime Oman! I will be waiting!"

Growth and Collaboration

Following Doring's defeat of Pablo "Ambong" Alicante, Pedrito "Pedring" Romo, and Roman "Oman" Ladaño, the Doce Pares Club flourished and became a household name synonymous with eskrima throughout the Philippine Islands. Membership increased as did local support of the club. The Doce Pares Club became a melting pot of talented eskrimadors of many

---

[18] Cages used to contain cockfighting roosters.

different backgrounds. Many of the members specialized in various aspects of eskrima such as *daga* fighting, *solo olisi*, *doble olisi*, *punta y daga*, use of the *sibat,* and *bangkaw,* as well as different empty-hand arts such as *dumog,* boxing, and wrestling. Each would openly share their knowledge of eskrima with other members and collectively grow and improve. This productive sharing continued for many years until the club was unavoidably disbanded after the Japanese invasion of Cebu and the beginning of World War II.

CHAPTER 4
World War II

As a child, I was too young to understand the politics of war. Still, I remember the fear my family had and concern the Empire of Japan would soon invade the Philippine Islands. As Japan continued its aggressive military campaign across Asia, rumors of an impending Japanese invasion began to spread as my family prepared for the inevitable attack.

By the time I was born on June 10, 1938, in the barangay of Mambaling in Cebu City, the Empire of Japan was already at war with the Republic of China as part of the Second Sino-Japanese War[19]. The political stability around the world was getting worse. That following year, Nazi Germany invaded Poland[20], which resulted in France and the United Kingdom declaring war on Germany[21]. Germany went on to conquer most of Europe and allied with Italy and the Empire of Japan[22]. By June 1941, the allied forces of Germany launched an invasion of the Soviet Union, which engulfed most of Europe and Asia in war. In retaliation, the United States increased support to China and its European allies and placed economic sanctions on Japan. These sanctions included a ban on shipments of aviation fuel, scrap iron, steel, and gasoline. By this time, Japan had already begun its expansion throughout Southeast Asia and the South Pacific. However, the economic sanctions placed on them by the U.S. forced Japan to expedite control of the region and seize Indochina and the resource-rich islands of the Dutch East

---

[19] The Second Sino-Japanese War was a military conflict fought primarily between the Republic of China and the Empire of Japan from July 7, 1937 to September 2, 1945.

[20] The Invasion of Poland by Germany on September 1, 1939 marked the beginning of World War II.

[21] The Declaration of War by France and the United Kingdom was given on September 3, 1939, after German forces invaded Poland.

[22] The Axis Powers, was an alliance created by Germany, Italy and Japan during World War II.

Indies[23].

Not wasting time, Japanese military forces advanced rapidly across the region in a well-coordinated military campaign. But between Japan and the resource-rich islands of the Dutch East Indies were the Philippine Islands. The islands formed a natural barrier between Japan and the resources of Indochina and the Dutch East Indies. The islands were of strategic military importance in the region. They were crucial to Japan's plan to control the South Pacific. However, the Philippine Islands and much of the South Pacific were under the control of General Douglas MacArthur.[24] He had been recalled back to active duty on July 26, 1941, in response to the escalating conflicts throughout Asia and assigned to establish and command the United States Armed Forces Far East (USAFFE) in Manila, Philippines.

Japanese Invasion of the Philippines

Ten hours after the surprise attack on US Naval Forces at Pearl Harbor, Hawaii, the Empire of Japan launched a surprise attack on the Philippine Islands on December 8, 1941. In a series of airstrikes, Japanese aircraft destroyed over half of the Far East Air Force (FEAF) aircraft at Clark and Iba Fields in Northern Luzon. Simultaneously, a Japanese landing force made an unopposed landing at Batan Island[25] in the Luzon Strait and seized control of the airstrip on the island.

As a result of the surprise attack on Pearl Harbor, Hawaii, on December 8. 1941, the United States declared war on the Empire of Japan. Undeterred, the Japanese conducted air strikes on Del Carmen Field near Clark, Nichols and Nielson Fields near Manila and the US Naval Facility at the Cavite Naval Yard on December 10, 1941. Due to the damage, the Japanese inflicted on the airfields and fearing the U.S. naval fleet would no longer be protected, Admiral Thomas C. Hart[26], Commander of the U.S.

---

[23] The Dutch East Indies was an island archipelago and Dutch colony in the South Pacific that later became known as the country of Indonesia.

[24] General Douglas MacArthur (January 26, 1880 – April 5, 1964) received the Congressional Medal of Honor for his service in the Philippines Campaign and was the only general who conferred the rank of Field Marshal in the Philippine Army.

[25] Batan Island is the second largest island of the Batanes Islands north of Luzon, Philippines.

[26] Admiral Thomas C. Hart (June 12, 1877 – July 4, 1971) served during the Spanish-American War and World War II after which he served as a United States Senator for the State of Connecticut.

The forward magazine of the USS Shaw explodes during the Japanese attack on Pearl Harbor, Hawaii, December 7, 1941. (U.S. National Archives and Records Administration).

Asiatic Fleet, ordered his fleet to withdrawal from the Philippines. That same day, Japanese landing forces landed at Camiguin Island and several locations along the northern coast of Luzon to include Aparri, Gonzaga, and Vigan. On December 12. 1941, a landing force of 2,500 Japanese soldiers landed at Legazpi on the southernmost point of Luzon, followed by an amphibious landing on Mindanao near Davao a week later. By December 19, 1941, Japanese Forces landed on Mindanao near Davao City, Philippines. Then, on December 22, 1941, the main Japanese invasion of the Philippine Islands began. Japanese troops under the command of General Masaharu Homma[27] landed at three separate locations along the Lingayen Gulf on the west coast of Luzon, northwest of Manila. Japanese forces seized the initiative and landed a second wave of Japanese troops at Lemon Bay, in the southern Luzon and began their advance inland toward Manila.

---

[27] General Masaharu Homma (November 27, 1887 – April 3, 1946) was convicted of war crimes and executed by firing squad on April 3, 1946.

General MacArthur ordered General Jonathan Wainwright[28], the Commander of the North Luzon Force, to hold the main Japanese assault at Lingayen and keep the road leading to the Bataan Peninsula open for use by the South Luzon Force Commander, Major General George Parker[29]. He further ordered the remaining U.S. and Filipino forces on Luzon to withdraw to defensive positions on the Bataan Peninsula. At the same time, Japanese troops advanced past the beaches and accomplished most of their objectives by the end of the day. They were now in a position to move into central Luzon.

1st Company, 7th Tank Regiment of the Japanese Imperial Army advance toward Manila, Philippines, c. 1941.

The American and Filipino troops on Luzon consisted of roughly 31,000 personnel comprised of various regular Army, National Guard, Constabulary, and Commonwealth units. The majority of them lacked combat experience and the resources needed to defend against a massive Japanese invasion. American and Filipino troops made every effort to protect Luzon against the combat-experienced Japanese forces. However, there was little hope of overcoming the Japanese attack.

General Wainwright radioed General MacArthur in Manila and informed him that further defense of the Lingayen beaches would be "impracticable" against the Japanese forces advancing toward Manila. General MacArthur elected against a counterattack using the remains of his reserve force and, on December 24, 1941, ordered all American and Filipino troops to withdraw to defensive positions on the Bataan Peninsula. Almost immediately, USAFEE personnel began preparing for evacuation from

---

[28] General Jonathan M. Wainwright IV (August 23, 1883 – September 2, 1953) was the highest-ranking prisoner of war and was held as a POW by the Japanese until his liberation in August 1945.

[29] Major General George M. Parker, Jr. (April 17, 1889 – October 25, 1968) was held as a POW by the Japanese until his liberation in August 1945.

General Douglas MacArthur's escape from the Philippines on March 11, 1942 onboard four U.S. Navy PT (Patrol Torpedo) boats. Shown for illustrative purposes is PT-105. (U.S. National Archives and Records Administration).

Manila and relocating MacArthur's headquarters to the heavily fortified island of Corregidor offshore in Manila Bay. The next day, General MacArthur and Manuel Quezon[30], the President of the Commonwealth of the Philippines, were evacuated from Manila to the island of Corregidor. By December 26, 1941, all USAFFE personnel had evacuated the city, and MacArthur issued a proclamation officially declaring Manila an Open City[31].

In January 1942, believing they had secured Luzon, Japan withdrew many of their forces and the bulk of their airpower from Luzon. This was done so by the Japanese to support the Japanese campaign in Borneo and

---

[30] Manuel Luis Quezon y Molina (August 19, 1878 – August 1, 1944) served as president of the Commonwealth of the Philippines from 1935 to 1944 and was considered the second president of the Philippines.

[31] An Open City is a city that is officially declared demilitarized during a war giving it immunity from military attack and allowing it to be safely occupied by its residents.

Indonesia. The withdrawal enabled the American and Filipino troops defending the Bataan Peninsula to successfully persevere against the Japanese forces left behind and endure for the next several months. Realizing their mistake, Japan swiftly strengthened their attacks and deployed reinforcements against the American and Filipino troops defending the Bataan Peninsula.

Fearing that Corregidor would soon fall to the Japanese and MacArthur would be taken prisoner, the United States President Theodore Roosevelt ordered MacArthur to evacuate the Philippines on March 11, 1942, and relocate to Australia. The next day, General MacArthur delegated command of all US and Filipino forces to General Wainwright and departed Corregidor for Mindanao with several members of his family and his Chief of Staff, Major General Sutherland,[32] onboard four US Navy PT boats commanded by Lieutenant Commander John D. Bulkeley[33]. From Mindanao, MacArthur flew to Australia onboard a B-17[34] bomber. Arriving at Batchelor Field in the Northern Territory of Australia on March 17, 1942, MacArthur flew to Alice Springs, where he was transported through the Australian outback by passenger train to Adelaide.

General MacArthur's retreat from the Philippines left him feeling frustrated and angry. In doing so, he left behind 90,000 American and Filipino troops at Corregidor and the Bataan Peninsula. They lacked food, supplies, and adequate support to defend against the amassing Japanese forces. In a statement to the press from a railway stop in the small township of Terowie, South Australia, on March 20, 1942, MacArthur promised his men and the people of the Philippines, "I shall return."

Following his arrival in Adelaide, MacArthur relocated to Melbourne, where he arrived on March 21, 1942. MacArthur immediately met with Allied military commanders and established the South West Pacific Area (SWPA) Command on April 18, 1942. Primarily consisting of American and Australian forces, the SWPA served as one of four major Allied commands

---

[32] Major General Richard Sutherland (27 November 1893 – 25 June 1966) served as General Douglas MacArthur's Chief of Staff in the South West Pacific Area during World War II.

[33] John Duncan Bulkeley (19 August 1911 – 6 April 1996) was one of the most decorated naval officers in US history. He received the Medal of Honor for actions in the Pacific Theater during World War II.

[34] A Boeing B-17 is a large four engine heavy bomber developed in the 1930s for the United States Air Corps that is commonly referred to as the Flying Fortress.

in the Pacific War and encompassed the Philippines, Borneo, the Dutch East Indies, East Timor, Australia, the Territories of Papua and New Guinea, and the western part of the Solomon Islands.

Surrender of Bataan

On March 28, 1942, Japanese forces began launching a wave of air and artillery attacks on the remaining Allied troops on the Bataan Peninsula. They were severely weakened by malnutrition, sickness, and prolonged fighting. Finally, on April 3, 1942, Japanese forces launched their final assault against the courageous American and Filipino forces defending the Bataan Peninsula. Within days, the Japanese troops were able to contain the remaining Allied troops on the peninsula and capture the observation post and artillery station at Mount Samat.

The "Bataan Death March" after the surrender of the Bataan Peninsula to Japanese Forces on April 9, 1942. (U.S. National Archives and Records Administration).

After four months of fighting and suffering heavy losses on Luzon, the remaining American and Filipino troops were exhausted, sick, and starving. Major General Edward P. King[35] feared his men would be overwhelmed by the advancing Japanese forces. He disobeyed direct orders from Wainwright and MacArthur to counterattack. Instead, he surrendered the Bataan Peninsula to the Japanese on April 9, 1942. The surrender was the largest surrender of a military force in United States military history.

The surrender of the Bataan Peninsula led to the transfer of all US and Filipino prisoners of war from Saisaih Point and Mariveles to a POW camp at Camp O'Donnell in the province of Tarlac. The Japanese forced the POWs

---

[35] Major General Edward P. King Jr. (July 4, 1884 – August 31, 1958) was held as a POW by the Japanese until his liberation in August 1945.

to march from Saisaih Point and Mariveles to the city of San Fernando in the province of Pampanga approximately 70 miles away. Once in San Fernando, the prisoners were packed into boxcars and transported by train to the Capas Train Station. They were then offloaded and forced to march the remaining nine miles to the POW camp at Camp O'Donnell. Prisoners were subjected to severe physical abuse and malicious killings as they walked in the scorching heat. The march resulted in the death of an estimated 1,000 American soldiers and 9,000 Filipinos and later became known as the "Bataan Death March."

Japanese Invasion of Cebu

For days leading up to the Japanese invasion of Cebu, air raid sirens frequently warned of Japanese planes as they regularly flew over the city. Each time we would evacuate to bomb shelters or take whatever cover was available as we prepared for the inevitable arrival of the Japanese forces. We would stay inside the bomb shelters until we received the "All clear!" then would nervously come out and go about our daily activities. Finally, on Friday, April 10, 1942, approximately 12,000 Japanese troops landed on the island of Cebu from eleven naval transport carriers at seven separate locations along the coastline. The invading Japanese forces encountered very little resistance as they occupied the island. Severely outnumbered and not supplied with the weapons and equipment needed to defend themselves, the American and Filipino troops on the island could not fight against the invading Japanese forces. Tragically, they were forced to abandon their positions and retreat into the mountains surrounding the city, along with thousands of Filipino civilians.

I remember watching the Japanese soldiers marching into Cebu City and the fear and panic that ensued. The Japanese were ruthless soldiers. They surrounded the city and rounded up all government officials and police and threatened to kill them and their families if they failed to cooperate. Many were ordered to be *undercovers*[36] for the Japanese to root out those not obeying their orders or joining the guerilla resistance so they could be captured and punished. Unbelievably, many traitorous Filipinos seeking immediate benefit and safety willingly supported the Japanese. They

---

[36] Undercovers were Filipinos who deliberately chose or were persuaded to become spies and informants for the Japanese military in exchange for preferential treatment and financial gain.

volunteered to be *undercovers* so they could receive preferential treatment and reward payments from the Japanese.

The Tribune, Manila, Philippines, April 24, 1942.

The Japanese military began controlling all public transportation, and sections of the city were cordoned off to monitor foot traffic. Japanese soldiers conducted house-to-house searches and immediately confiscated any firearms or weapons found as were any radios or communication devices. The Japanese military established strict control of all public communication and took control of the local radio stations. Resisting or disobeying any orders was met by severe punishment that often resulted in torture, death by beheading, firing squad, or being stabbed to death by bayonets.

During the invasion, my dad Vicente "Inting" Atillo, Delfin Lopez, and Teodoro "Doring" Saavedra escaped and fled to the safety of the mountains to join the fledgling guerilla resistance movement. Venancio "Anciong" Bacon fled south to be with his family in Carcar,[37] where he remained until the war was over. I stayed with my mother and family in Mambaling, as did Lorenzo "Tatay Ensong" Saavedra, who was 90 years old at the time of the Japanese occupation. Most women and children were marginally safe from the occupying Japanese forces if they showed no

---

[37] The City of Carcar is a small coastal city on the east coast of Cebu approximately 25 miles south of Cebu City, Philippines.

resistance and immediately complied with their orders and demands. This safety was not always the case, however, and many women and children received horrific treatment and abuse. Hundreds of women were raped, tortured, and killed, and the Japanese often abused their children. Sometimes even killing them for no reason. My dad, Delfin Lopez, and Doring Saavedra opposed the Japanese occupation, and they would have been immediately killed if captured during the invasion. Anyone suspected of supporting or being a member of the fledgling guerilla resistance was quickly arrested and taken to the newly established Kempeitai Headquarters (HQ) at the Cebu Normal School in Cebu City and subjected to cruel punishment and death. The *Kempeitai*[38] were the military police branch of the Imperial Japanese Army and were notorious for their brutality and cruelty.

The Capture of Corregidor and Surrender to the Japanese

Following the surrender of the Bataan Peninsula, Japanese forces focused their efforts on the island of Corregidor, which protected the entrance to Manila Bay. Corregidor was the last remaining obstacle for the Japanese, and capture of the island would undoubtedly lead to a complete surrender of the Philippines to the Empire of Japan.

From his SWPA headquarters in Melbourne, Australia, General MacArthur continued his support of General Wainwright's efforts to defend the Philippines. However, defense of the islands was becoming increasingly improbable. Shortly after midnight on May 5, 1942, after an intense barrage of shelling, Japanese forces landed on Corregidor with additional reinforcements landing throughout the night. General Wainwright and the remaining American and Filipino troops on Corregidor courageously defended the island in a massive battle throughout the night. Unfortunately, the increase in Japanese soldiers and the severe lack of weapons, ammunition, food, and supplies placed overwhelming stress on the enduring forces on the island. Recognizing the inevitable, Wainwright reluctantly asked Lieutenant

---

[38] The Kempeitai were a military police branch of the Imperial Japanese Army that served as both a conventional and secret military police force. They were known for their brutality, callousness, and brutality and were responsible for the rape, torture and death of hundreds of Filipino citizens during World War II.

General Masaharu Homma[39], commander of the Japanese 14th Army, for his terms of surrender on May 6, 1942.

Homma's reply was clear and direct, "Surrender everything or nothing." If he refused, Homma advised Wainwright he would execute everyone on Corregidor. Wainwright recognized he was facing impossible odds, and there was no other feasible option. Consequently, Wainwright instructed Brigadier General Lewis C. Beebe, Assistant Chief of Staff of U.S. Forces, to broadcast over the "Voice of Freedom[40]" radio a message to Homma accepting his terms and the surrender of all American and Allied forces in the Philippines.

Almost concurrently with Beebe's broadcast, Wainwright instructed his radio operator to send a coded radio message to Brigadier General William Sharp[41], Commander of the Visayan-Mindanao Force. Wainwright relinquishing command of all remaining American and Allied forces in the Philippines to Sharp except Corregidor and the islands in Manila Bay and instructed him to report to General MacArthur immediately for orders.

General Wainwright broadcasting instructions to surrender over Station KZRH, May 7, 1942. (Naval History and Heritage Command, Washington, DC).

The purpose of Wainwright's decision to relinquish command to Sharp was to surrender as few men as possible. Wainwright hoped to convince Homma

---

[39] Masaharu Homma (November 27, 1887 – April 3, 1946) commanded the 14th Army of the Imperial Japanese Army during World War II and carried out the Bataan Death March. Homma was convicted of war crimes and was executed by firing squad on April 3, 1946.

[40] The Voice of Freedom was a radio broadcast established on January 5, 1942 in the Malinta Tunnel on the island of Corregidor. The radio severed as a means of broadcasting information from the underground USAFFE headquarters of General Douglas MacArthur.

[41] Major General William Sharp (September 22, 1885 – March 30, 1947) surrendered his command to the Japanese after the Fall of Corregidor and spent the rest of the war as a Japanese POW.

that the Allied forces in the south were not under his control, and he did not have the authority to order their surrender. Unfortunately, Homma refused to believe that Wainwright was no longer the commander of all forces in the Philippines. He instructed Wainwright to surrender all troops to the local commander at Corregidor, Colonel Gempachi Sato, Commander of the 61st Infantry Division. Just before midnight on May 6, 1942, General Wainwright met with Colonel Sato and officially signed the surrender documents that contained all provisions Homma insisted upon. By signing, Wainwright unconditionally surrendered the Philippines to the Empire of Japan and instructing all local commanders to assemble their troops and surrender to the nearest Japanese commander.

Sharp knew that if he ignored or refused Wainwright's orders to surrender, Homma would execute Wainwright and massacre all military personnel and civilians at Corregidor. He also recognized that many troops under his command in the region would most likely refuse to surrender. That would continue fighting against the Japanese as guerrillas—particularly the Filipinos who made up the vast majority of his command. Sharp had already designed plans to withdrawal to the more remote areas of the island to establish organized resistance groups. His local commanders were ready to execute them once the order was received from General Sharp.

Sharp relayed the message received from Wainwright to General Douglas MacArthur in Melbourne, Australia, and asked for further instructions. MacArthur's instructions were quickly received. Sharp was ordered to separate his force into small elements and initiate guerrilla operations. Sharp had been given complete authority to act on his judgment by MacArthur but struggled with the order to surrender previously given by Wainwright. He knew that refusing would lead to the massacre of everyone at Corregidor. To buy time, Sharp elected to wait for the arrival of Wainwright's messenger, Colonel Jesse Traywick, Sr. and Colonel Hikaru Haba before making his decision. They had been flown to Mindanao to relay Wainwright's instructions to surrender. However, Sharp acted upon MacArthur's order and released the island commanders under his command and instructed them to prepare for guerrilla operations.

Traywick and Haba arrived on Mindanao on May 9, 1942, and immediately arranged a meeting with Sharp. Traywick delivered Wainwright's letter to Sharp and explained the circumstances that led to Wainwright's decision to surrender. He made clear that if the Visayan-

Mindanao Force refused Wainwright's order, the Japanese would reject the terms already agreed upon and would likely execute everyone on Corregidor. Fearing that Homma would follow through with his threat to kill everyone on Corregidor, Sharp reluctantly surrendered on May 10, 1942. However, he knew his commanders were already preparing to conduct guerrilla operations against the Japanese.

Colonel Bradford Chynoweth[42], Commander of the 61st Philippine Division on the Island of Cebu, was also resistant to follow Wainwright's order to surrender. He had already made extensive preparations for guerilla operations on Cebu and felt the complete surrender of American and Filipino forces was unnecessary. Chynoweth had heard the radio broadcast to surrender. However, he believed that Wainwright's order was likely done so under duress and may have been a ruse by the Japanese. Chynoweth relayed Wainwright's order to surrender to the various units under his command however told his men they could "surrender individually if they wished to do so." The ambiguous order encouraged many of the outlining groups to evade capture and established the foundation for the future guerilla resistance on Cebu. On May 15, 1942, Chynoweth received the orders sent to him by General Sharp, confirming Wainwright's order to surrender. Reluctantly, Chynoweth assembled the remaining elements of his force and surrendered the island of Cebu to the Japanese the following day on May 16, 1942.

James Cushing and the Cebuano Guerillas

One member of Chynoweth's command who refused to surrender was a young Army captain with the USAFEE Corps of Engineers named Captain James M. Cushing[43]. Before the war, Cushing was an American mining engineer of Mexican-American descent. He relocated to the Philippines with his brothers,

---

[42] Colonel Bradford Chynoweth (July 20, 1890 – February 8, 1985) laid the foundation for the guerrilla campaign in the Visayan Islands prior to the surrender of the Philippines to the Empire of Japan.

[43] Lt. Col. James M. Cushing (b. circa 1910 - August 26, 1963) was instrumental in the guerilla campaign against the Japanese military and was instrumental in the infamous Koga affair in which the Z Plan of the Imperial Japanese Navy was recovered by his guerillas. Cushing survived the war however died of a heart attack on August 26, 1963 on an inter-island transport to Mindoro Island. He is interned in the Hero's Cemetery in Manila, Philippines.

Walter Cushing[44] and Charles J. Cushing,[45] to take advantage of the booming mining industry. After visiting several islands throughout the Philippines, Cushing met and married Felisa "Fritzie" Tabando from Leyte and settled in Cebu.

Cushing was a fighter who had no fear whatsoever. Before the invasion of Cebu by the Japanese, Cushing wired explosive charges and destroyed most major bridges and railways on Cebu, so the Japanese could not use them. After the invasion, the Japanese were quick to blame the fledgling guerilla resistance for the damage; however, it was Cushing himself who was responsible for the destruction. Instead of surrendering, Cushing gathered his Filipina wife Fritzie, his dog Senta, a Great Dane, and escaped into the mountainous interior of Cebu.

Pre-World War II photograph of Lt. Col. James Cushing (1910 – August 26, 1963).

It didn't take long for the men who evaded capture to organize small guerilla resistance groups throughout the island. One of the first to formally organize a group of resistance fighters was Lieutenant José Macabuhay from Mambaling. With the support of the governor of Cebu, Hilario "Dodong" Abellana, Macabuhay was able to secretly assemble a small group of patriots from my barangay of Mambaling in early July 1942. Some of the first to join

---

[44] Major Walter M. Cushing (August 12, 1907– September 19, 1942) became Commander of the Northern Luzon guerillas in Ilocos Norte, Philippines. During a Japanese ambush, instead of being taken alive, Cushing killed himself on September 19, 1942 with the remaining round of his .45 caliber pistol. Impressed by his bravery and honor, he was given a proper burial by the Japanese military and buried in the Municipality of Jones, Province of Isabela in Northern Luzon, Philippines. On November 4, 1949, his remains were disinterred and brought back to the United States and reburied at Fort Rosecrans National Cemetery in San Diego.

[45] Major Charles J. Cushing (b. circa 1919 – d. unknown) became Commander of the Pangasinan Guerilla Force in Northern Luzon, Philippines. He surrendered to the Japanese military after his wife Mercedes was captured in March 1943.

was my dad, Teodoro "Doring" Saavedra, and Delfin Lopez. Others included Evaristo Abayan, Celmente Alda, Max Alda, Catalino Andaya, Severino Bermejo, Guillermo Gadiana, Bonifacio Gadiana, Miguel Paez, Perferio Paez, Nono Rosales, and Felipe Torres. In addition, many other eskrimadors joined the group, including Doring's brother Frederico "Pedring" Saavedra. He was also a childhood student of Lorenzo "Tatay Ensong" Saavedra and a member of the Doce Pares Club before the war.

Walter M. Cushing (d. September 19, 1942)

Abellana believed that the only chance the Filipinos had against the Japanese was to slowly and methodically form an organized guerilla resistance movement against the Japanese military. Abellana believed that once the Japanese military settled in Cebu, most of them would be reassigned to other areas in the region to support the war in the Pacific. The number of soldiers on Cebu would be reduced to a minimum. Abellana believed it would be best to gather and stockpile weapons and munitions, then attack the Japanese after troop levels on the island had been downsized. He then instructed Macabuhay and his small group of guerillas to collect firearms, ammunition, explosives, mortars, and any other resources they could find and hide everything in the abandoned mine tunnels outside the city. The newly formed guerilla group also began conducting small reconnaissance missions and harassment raids against the Japanese as early as July 1942. On one such mission, the group successfully infiltrated into Cebu City and stole four Nambu[46] pistols, sixty rounds of ammunition, two samurai swords, three Japanese rifles with ammunition, and bayonets from the Japanese garrison in Basak.

Although other guerilla groups were being formed throughout the

---

[46] A Nambu Pistol is a series of semi-automatic pistols manufactured by the Koishikawak Arsenal in Japan from 1906 - 1942.

Filipino guerillas practicing close quarters combat and the art of eskrima during World War II, Philippines. (U.S. National Archives and Records Administration).

Philippine Islands to fight against the Japanese, they would not be effective unless they were united together as a cohesive fighting force and lead by a strong leader who could be trusted. Also, not all of these groups were guerillas who shared the same desire to fight against the Japanese and liberate the Philippine Islands. Many were criminals, thugs, and bandits who formed gangs to enrich themselves and rob innocent civilians. Some even became undercovers for the Japanese military to receive preferential treatment and payment for spying on other Filipinos and providing information to the Japanese. Politicians, mayors, and barangay captains in the southern part of Cebu began pressuring Cushing into heading up the guerilla effort and bring them together as a united front. As a captain with the USAFEE Corps of Engineers, he had earned a reputation as a fearless leader that the Filipinos trusted and respected. Finally, in August of 1942, Cushing agreed and began efforts to unite the various guerilla groups operating independently throughout the island of Cebu into a single fighting force.

A short time later, Cushing met with Harry Fenton,[47] an American radio broadcaster in Cebu City who fled to the barangay of Tabunan with his Filipina wife, Betsy, and their baby son, Steve, during the Japanese invasion. Fenton despised the Japanese and had already been organizing a resistance group in the densely forested mountains of Tabunan. Cushing and Fenton agreed to unite and consolidate the existing guerilla groups on Cebu into a single joint command called the Cebu Area Command (CAC). They further separated their roles and responsibilities into two distinct functions. Cushing would handle all training and combat operations in the field, and Fenton would manage administration. Cushing and Fenton further decided that the headquarters of the new CAC would be at Fenton's isolated camp in the barangay of Tabunan, a short distance from Cebu City.

In August 1942, my dad, Doring Saavedra and Delfin Lopez conducted a demonstration of boxing, wrestling, and eskrima for Cushing at the newly formed CAC headquarters at Tabunan. Cushing was very impressed by the eskrimadors and immediately saw the value of their hand-to-hand combat skills. Many guerillas were civilians who had no military training, and only a handful were former USAFFE soldiers. Cushing asked them to join the CAC as well as teach eskrima and hand-to-hand combat to the other Cebuano guerillas. They agreed and were formally inducted into USAFFE by Capt. James Cushing on August 15, 1942. Along with Doring Saavedra and Delfin Lopez, my dad's first assignment was to the Combat Company of the 85th Infantry Regiment commanded by Lt. Rogaciano "Popoy" C. Espiritu and Lt. Macabuhay.

Under orders from Cushing, the three eskrimadors immediately began teaching eskrima and hand-to-hand combat to their fellow guerilla fighters. Their orders were to spread their skills to the outlining guerilla units and show the other guerillas how to kill Japanese soldiers in close combat. A short time later, they were able to add nine additional eskrimadors to the cadre of instructors and began to refer to themselves as the Doce Pares of Tabunan.[48]

---

[47] Lt. Col. Harry Fenton (b. circa 1907 – September 16, 1942) also known Aaron Feinstein, was a former member of the US Army Medical Corp. He was executed by firing squad by his own men for violation of several provisions of the Articles of War (AOW).

[48] The number of instructors selected as members of the Doce Pares Club of Tabunan was a coincidence, however it symbolized the loyalty and respect the eskrimadors had for Lorenzo "Tatay Ensong" Saavedra and the original pre-war Doce Pares Club.

This group of young patriotic eskrimadors included Filemon "Moni" Caburnay[49], who credits learning *punta y daga* from Doring during his time as a member of the unit.

Assassination Mission

On August 30, 1942, my dad and the newly formed Cebuano guerilla unit received orders to kill a Filipino undercover working for the Japanese named Mariano T. Jaucian. Jaucian had been recruited by the Japanese Kempeitai as a spy to gather intelligence and identify those involved in the underground guerilla movement. The Kempeitai was a ruthless military police branch of the Imperial Japanese Army and were renowned for their savagery. When Jaucian had been working for the Japanese, he had aided in the capture, torture, and execution of several Filipinos. He had even gone on raids with the Japanese to capture members of the underground guerilla resistance.

Filipino guerrillas assault a village hut, Leyte island, Philippines, November 1944. (U.S. National Archives and Records Administration).

Jaucian's location was unknown until earlier that day when a member of the guerilla unit located the house where Jaucian lived in the barrio of Tisa, west of Cebu City. The guerillas put a hasty plan together to assassinate Jaucian while he was sleeping and began their preparations. Armed with aged Springfield

---

[49] Filemon "Moni" Caburnay combined the punta y daga techniques he learned from Teodoro "Doring" Saavedra with the Arnis de Abanico he learned from his brother Arsenio "Siniong" Caburnay (1882-1962) and Artemio "Timyong" Paez, and what he learned from Filemon "Momoy" Cañete, and founded the Lapunti Self Defense Club on December 30, 1972.

M1903 rifles and the Nambu pistols captured from the Japanese garrison at Basak, the guerillas departed Tabunan at midnight to conduct the secret assassination mission.

As the guerrillas approached Jaucian's residence and started maneuvering into position, a dog sensed their presence and began barking, compromising their assault. Without delay, the guerillas commenced firing into the house, focusing their gunfire toward the room where they believed Jaucian was sleeping. A flurry of bullets perforated the house and riddled the room as the occupants inside screamed and scrambled for cover. The barrage of gunfire alerted a group of Japanese soldiers standing nearby who swiftly assembled a hasty response force. In the commotion and confusion, the guerillas retreated into the darkness and safety of the mountains just as three trucks loads of Japanese soldiers arrived.

The next day, the guerillas found out that during the assault, Jaucian had been sleeping in another room with his *kerida*[50] and survived the assassination attempt. Sadly, one of the bullets hit Jaucian's wife in the leg, and a child inside the residence later died from injuries sustained during the assault. The assassination mission was unsuccessful. However, it communicated clearly to Mariano T. Jaucian that his secret life as a Japanese undercover was compromised, and the guerillas had placed a price on his head. More importantly, the mission demonstrated the unpredictability of the guerillas and proved to the Japanese military that the guerillas could attack anywhere in Cebu City without warning.

Battle of Babag

The new guerilla unit engaged in their first major battle against Japanese forces on October 24, 1942. Lieutenant Rogaciano "Popoy" Espiritu and Sergeant Ramon Climaco led two platoons to Lahug on the outskirts of Cebu City. Including my dad, Doring Saavedra, and Delfin Lopez. Espiritu ordered his platoon to assault the University of the Philippines Cebu College to rescue several captive American prisoners simultaneously as the second platoon attacked the Lahug Elementary School to rescue captive British citizens. The remaining men would deploy as a blocking force at the junction leading to Lahug Field, a military airfield occupied by the Japanese Army Air Force.

---

[50] A *kerida* is a Tagalog term for a mistress adopted from the Spanish word *querida*.

As the platoons separated and began maneuvering into position, the men guarding the junction leading to Lahug Field identified a column of Japanese soldiers moving toward their position. The men took cover and ambushed the Japanese. The first barrage of gunfire from the blocking force killed thirty Japanese soldiers. However, unexpected contact with the Japanese compromised the mission. It forced the guerillas to withdraw to the safety of the mountains and regroup.

The guerillas suffered no casualties, but the surprise attack left the Japanese frustrated and angry. On October 26, 1942, the Japanese assembled a force of 400 soldiers and launched a counter-offensive into the mountains. The Japanese wanted to root out the guerillas responsible for killing their men a few days earlier. As they approached the hills of Tagid and Paypay, the Japanese encountered Lt. Espiritu and his men who were vastly more experienced and knowledgeable in the challenging, dense, jungle terrain and forested mountains outside Cebu City. A massive firefight ensued that lasted for several days.

As the battle raged on, the Japanese requested air-drops to resupply their soldiers with food and ammunition. A short time later, several Japanese planes appeared overhead that began dropping cases of supplies for the Japanese soldiers. Unfortunately for the Japanese, wind conditions in the area were unpredictable, and many of the small parachutes drifted off course into areas controlled by the guerillas. The guerillas observed the misdirection of the chutes. They were able to recover the supplies before the Japanese were able to secure them.

The battle between the Japanese soldiers and the Cebuano guerillas of the Combat Company of the 85th Infantry Regiment continued for days. Finally, realizing they would not get the upper hand on the guerillas in the challenging terrain, the Japanese soldiers withdrew. The Cebuano guerillas suffered no losses, but the battle took a heavy toll on the Japanese.

Train Assault at Inayawan Crossing

By December of 1942, Captain Cushing had begun focusing his efforts on the railway that runs north and south between Danao and Argao along the eastern coast of Cebu. The Japanese military regularly used the railway to transport personnel and supplies; however, they were also very aware of the likelihood of guerilla attacks along the tracks.

On December 28, 1942, the Cebuano guerillas received information that a train loaded with Japanese soldiers would transit from Carcar to Cebu City. Cushing ordered Lt. Nestor Legaspi and his guerillas to ambush the train where the railroad tracks crossed the provincial road from the south to Cebu City. Expecting a train loaded with Japanese soldiers, Legaspi requested reinforcements from Captain Cushing, who sent Lt. Jesus Navarro and Lt José Macabuhay. Among Macabuhay's men were my dad, Doring Saavedra, and Delfin Lopez.

As the train approached the ambush site, Legaspi noticed that not only was the train fully loaded with Japanese soldiers, there was a column of soldiers marching alongside on the eastern seaside of the train. The Japanese were expecting an ambush by the guerillas. They were using the train to hide and protect the column of soldiers marching alongside. To further protect the train, the Japanese intermittently launched mortars from the train toward the mountains along the tracks as the train gradually moved through the area. As the train reached the crossing, the guerillas opened fire. The train screeched to a stop as the Japanese foot soldiers on the seaside of the train scrambled to find cover positions. As they did, the guerillas who were prepositioned to ambush the train on the eastern shoreline attacked, leaving the Japanese soldiers with little chance to survive. In the wake of the short but intense firefight, 180 Japanese soldiers lay dead, and none of the Cebuano guerillas died during the altercation.

Railroad Crossing, North Road, Danao, Cebu, Philippines. (U.S. National Archives and Records Administration).

The success of Captain Cushing and the Cebuano guerillas of the Cebu Area Command enraged the Japanese. The Japanese military constructed a bamboo perimeter fence surrounded the city to protect themselves from the unpredictable guerilla attacks and control access into Cebu City. Gates were placed at several locations around the city and guarded by Japanese soldiers who controlled access and searched everyone who exited or entered the city. One entrance was near my home in Mambaling at the end of C. Padilla Street, about 500 yards from the Kinalumsan Bridge. I remember watching hundreds of Japanese soldiers exiting through the gate to conduct operations against the Cebuano guerillas hiding in the mountains just outside Mambaling.

Farmhouse Ambush

Soon after the Japanese completed construction of the perimeter fence, approximately 200 Japanese soldiers exited the gate near Mambaling and advanced toward Buhisan Mountain. The soldiers stopped to rest at a farmhouse on the road to Barrio Tabok-Canal. Seizing the opportunity, Lt. Legaspi maneuvered his guerillas into positions around the farmhouse. At approximately 8:30 a.m., the guerillas opened fire. The battle raged on into the afternoon. Lt. Macabuhay intercepted Japanese reinforcements sent to aid the soldiers at the farmhouse, and his men acted as a blocking force between Mambaling and Cebu City.

The fight lasted well into the night. Only eleven Japanese soldiers were able to escape and make it back to the gate alive. Captain Cushing and the Cebuano guerillas outside Mambaling recovered more than 100 rifles, several rounds of ammunition, and mortars left behind by the retreating Japanese and the dead Japanese soldiers killed in the ambush. The ambush at the farmhouse was a significant victory for my dad and the Cebuano guerillas. They were running low on supplies and in desperate need of additional weapons and ammunition.

Death of an Eskrima Icon

By the end of 1943, the Japanese had become tremendously frustrated by the success of Captain Cushing and the Cebuano guerillas. Efforts by the Japanese to capture Cushing had been unsuccessful, and they were becoming desperate. The Japanese knew they could not effectively engage the guerillas in their native surroundings and the dense jungle and densely forested mountains outside Cebu City. As a last-ditch effort, the Japanese began using propaganda and deception tactics to lure guerillas away from the safety of the mountains and into the urban areas of the city. Once inside, they could be captured and killed.

In October 1943, the Japanese scattered propaganda posters throughout Cebu City and air-dropped leaflets along the densely forested mountains of Cebu advertising a "Day of Peace." The invitation offered a day of temporary amnesty to guerillas hiding in the mountains. The leaflets stated they could safely enter Cebu City and visit their family members without fear of being harassed or harmed. However, once inside the fenced perimeter surrounding the city, the Japanese had planned to capture, torture, and kill anyone suspected of being a guerilla.

Doring Saavedra had been hiding in the mountains since the beginning of the Japanese occupation and decided to accept the offer so he could enter the city and visit his family. My dad and Delfin Lopez explained to Doring that the proposal was a ploy by the Japanese, and he would indeed be captured and killed if he accepted the offer. Undaunted by the risk and determined to see his family, Doring explained to his *compadres* that this was his only chance to leave the mountains and visit his family in Mambaling. Against the desperate pleas of my dad and Delfin Lopez, Doring reassured his lifelong friends and compatriots he would return. He then proceeded to make his way to the outskirts of Cebu City.

As Doring approached the perimeter fence surrounding the city, the Japanese soldiers guarding the gate confronted and challenged him. Doring showed the guards the leaflet he had found advertising a "Day of Peace" and stated his command permitted him to enter the city to see his family. As the Japanese soldier's conferred with one another in Japanese, Doring began to feel that maybe my dad and Delfin Lopez were right. Perhaps this was all a ruse to draw the guerillas from the mountains so they could be tortured and killed. Surprisingly, the guards approved Doring's request to enter and ushered him through the gate and into Cebu City.

Doring cautiously walked away from the Japanese-guarded gate with a sense of relief and exhilaration as he continued walking toward his home in Mambaling. He was excited to see his family and friends, along with his 91-year-old uncle and famed eskrimador, Lorenzo "Tatay Ensong" Saavedra, whose health was beginning to decline. Doring continued walking when suddenly he was confronted by several soldiers of the feared Japanese Kempeitai.

Doring again showed the soldiers the leaflet he had found and stated he had been permitted to enter the city to see his family. Suddenly, the ranking Kempeitai soldier pointed at Doring and shouted in Japanese, "You are a guerilla! You are under arrest!" As the Japanese had deceitfully planned, the Kempeitai soldiers arrested Doring for being one of Cushing's guerillas. They ordered him to climb onboard a nearby military truck. Doring became angry and refused, at which time the ranking soldier angrily again shouted, "You are under arrest! Get on the truck!" in an attempt to get Doring to comply with his orders. Doring knew what would happen if he boarded the truck, so he again refused. The soldiers lunged forward and grabbed Doring in an attempt to force him to comply physically. Doring resisted and began fighting back against the soldiers who were no match against the combat-hardened eskrimador.

The soldiers abandoned their futile attempt to restrain Doring and pointed their rifles at him and shouted, "Get on the truck, or you will be shot!" Realizing he had no other choice, Doring reluctantly climbed on board the truck. He was then transported to the Japanese military garrison at the Basak Elementary School[51], a detention camp for Filipino prisoners of war.

Upon arrival, Doring was ordered to climb down from the truck and instructed to follow several Kempeitai soldiers who led him at gunpoint to a clearing inside the school between two adjacent buildings. The soldiers then tied Doring's hands and feet together with rope. They began beating him with a makeshift baseball bat and striking him repeatedly with their rifle-butts. Despite their efforts to interrogate him for information, Doring refused to tell them anything that would compromise Cushing and the Cebuano guerillas or the safety of his fellow compatriots. Doring's strong will infuriated the soldiers. They began beating him even more with merciless vigor to force him to provide information about his fellow guerillas. If not, he would die from the pain caused by the brutal interrogation and punishment.

Japanese Troops executing Chinese prisoners during the Sino-Japanese War (1937-1945). Live prisoners were often used by Japanese troops for bayonet practice. (U.S. National Archives and Records Administration).

---

[51] The Basak Elementary School was renamed Don Vicente Rama Memorial Elementary School and is located on Macopa St., Cebu City, Philippines.

Suddenly, Doring was able to free himself from the ropes binding his hands and feet together and attempted to escape. The soldiers struggled to regain control, but as one of Cebu's greatest eskrimadors, Doring could defend himself and fight back. Fighting for his life, Doring continued his attack as the soldiers feverishly struggled to regain control. Suddenly, one of the soldiers drew his samurai sword and attempted to slash Doring, who narrowly evaded the attack. Doring then began struggling with the sword-wielding attacker to control the blade. At that time, another soldier stabbed him from behind with a bayonet affixed to his rifle.

Bleeding profusely and facing impossible odds, Doring fearlessly and bravely continued his fight for survival. He began to overpower the Kempeitai soldiers with the little life he had left. The soldiers realized they were not going to regain physical control of the legendary eskrimador using their Japanese hand-to-hand combat techniques. Suddenly, one of the soldiers raised his rifle and shot Doring at point-blank range, killing him on the spot. Doring collapsed to the ground, dead, but only after he singlehandedly fought back against several combat-experienced Japanese Kempeitai soldiers armed with swords, bayonets, and rifles. He did so while bleeding profusely from the bayonet wound in his back and damages received during the lengthy torture and interrogation by the Kempeitai. His body was then dragged away and disposed of along with hundreds of other Filipinos. Their lives had also been taken due to the atrocities committed by the Japanese military throughout the war.

My neighbors, Lucas Degamo and his friend Juan witnessed Doring's torture and his attempt to escape. Degamo and Juan were also being detained at the Basak Elementary School when Doring was captured, tortured, and killed. Following their release from detention, Degamo and Juan often talked of the bravery Doring displayed and the effectiveness of his fighting style. His death illustrated his greatness and legendary ability as a fighter and one of Cebu's greatest eskrimadors. As well as his love for his family, loyalty to his compatriots, and his devotion to his country.

Capture and Escape from Death

A short time later, the Japanese Kempeitai, with traitorous Filipino undercovers, began conducting retaliatory raids in Mambaling. They wanted to capture other members of the Cebuano guerillas who may have entered the city and arrest and torture their family members as vengeful retribution. I was at home in our kitchen, shucking corn with my mother and brother when the soldiers raided our house. They immediately arrested us and dragged us outside, where they loaded us onto the back of military trucks alongside other civilian captives who the Japanese had detained during the raids. The Japanese then transported us a short distance away to an area near the chapel in Mambaling. Not far from the Basak Elementary School where Doring had been executed only a short time before.

Once we arrived, the Japanese dragged us from the trucks and separated us by gender and age, and placed us at separate locations where we were ordered to sit. No one was allowed to stand up or speak. The Japanese Kempeitai were very strict and would slap and kick anyone who didn't follow their orders. As we sat silently, more trucks arrived, and other captives were offloaded and separated into the different groups.

The Kempeitai then lined up and formed a firing line in front of us with machine guns mounted to the back of military trucks with armed soldiers standing alongside. On command from a nearby Japanese officer, they elevated the machine guns mounted to the vehicles and raised their rifles in preparation for the order to execute us all in a mass execution. Fearing certain death, we all sat motionless in disbelief. We began to cry as we waited for the execution order and hail of gunfire that would instantly end our lives. In the distance, we could hear gunfire and the screams of innocent Filipino victims. They were being slaughtered by the ruthless Kempeitai. Even as a young boy, I was aware of what was happening and could sense the fear and feeling of helplessness and desperation in the air.

Suddenly, a military jeep arrived carrying a high-ranking Japanese officer. I recognized the officer the moment he exited the vehicle. It was the feared and ruthless Captain Kasatoro Tsuruyama[52], an ugly flat-nosed man with a rectangular flat head and mustached face. He was a feared Japanese officer, and everyone was afraid of him, including his own men. His orders were acted upon instantaneously and without question. His subordinates feared him as much as the Filipinos. To Captain Tsuruyama, the Filipino people were helpless pygmies who were no longer human beings.

Remarkably, instead of ordering our execution, Captain Tsuruyama

---

[52] Capt. Kasatoro Tsuruyama was the feared head of the Japanese Kempeitai in Cebu. Tsuruyama was tried and convicted for the murder of two American prisoners in Cebu City by the International Military Tribunal for the Far East and executed by hanging on June 6, 1946.

instructed the Kempeitai soldiers to let the women and children go. The soldiers executed Tsuruyama's orders obediently and began screaming at us to stand up and leave the area quickly. Without a second thought, I promptly jumped to my feet and ran from the field. In the disorder and confusion of the fleeing survivors, I was able to locate my mother and brother, who had also survived. Once my mother was able to gather us together, we began our escape from the area to the safety of Buhisan Mountain. It was there, my dad and the other Cebuano guerillas were in hiding just outside Cebu City.

As we followed the Buhisan River into the mountains from Basak, we stumbled upon a mass of dead bodies at the base of the hills on the outskirts of Mambaling. The bodies had been callously dumped into the river by the Japanese military. I could see flies swarming around the deceased and could smell the rotting flesh of bodies decomposing beneath the victims murdered only moments before. I stared in horror as we hurriedly negotiated our way past the lifeless remains. Suddenly, in shock and disbelief, I recognized the dead body of Doring Saavedra. I pointed out Doring's lifeless body to my mother. She quickly restrained me in fear of attracting unwanted attention by the ruthless Japanese soldiers nearby. She hurriedly ushered me and my brother passed the bodies into the mountains above the river where the guerillas hiding outside the fenced perimeter surrounding Cebu City had been waiting. Under the armed protection of the guerillas, they led us along a trail into the densely forested mountains where we reunited with my dad and Delfin Lopez. I explained to them what happened at the Basak Elementary School and of finding Doring's body in the mass grave along the river. Grief-stricken and with sadness in his voice, my dad said they had learned of Doring's death from other guerillas

The location where Teodoro "Doring" Saavedra's body was found in the Buhisan River, Cebu, Philippines. The elderly resident currently living along the river stated there were "hundreds" of bodies disposed of at the location by the Japanese during the war.

Boy soldiers help carry ammunition and supplies for Filipino guerillas west of Cagayan Valley, Luzon, Philippines, July 6, 1945. (U.S. National Archives and Records Administration).

who observed his capture and execution at the school.

    We remained in the mountains for several days until it seemed the Japanese soldiers had returned to their regular routine, and the unstable conditions in the area had calmed. Once it was safe, my mother and brother returned to our home in Mambaling, and I stayed with Delfin Lopez and my dad at their guerila camp in the mountains. I assisted my dad and the Cebuano guerillas any way I could. As a young boy, I was limited in what I could do as a boy soldier. My father and Delfin Lopez did their best to keep me safe and out of danger as much as possible, given the volatile conditions of war. I would fetch supplies, weapons, and ammunition and would stand lookout for Japanese patrols. I would often accompany my dad on reconnaissance missions and stand guard as they gathered intelligence and prepared for their missions. I was also frequently asked to stand guard over captured Japanese prisoners. They were captured and brought back to the camp for interrogation. Armed with a *sundang*[53], I would poke and cut them if they became unruly and threaten to kill them if they resisted.

---

[53] A sundang is a type of knife similar to a small bolo or machete.

I had seen first-hand the brutality of the Japanese soldiers, and I had grown to hate them. My hatred became even more fueled after experiencing the mass executions at the Basak Elementary School first-hand and discovering Doring's body in the Buhisan River. My early childhood years were during the war, and the memory of Doring's dead body lying in the Buhisan River haunted me. It still does to this day. I loved and idolized Doring, and I was enchanted and mesmerizing when I watched him practice eskrima with Tatay Ensong and my dad. His death changed me. As a small boy growing up in a nation occupied by Japanese soldiers and besieged by war, I had become calloused. I was willing to kill anyone to protect myself and my family.

Doring's body remained in the Buhisan River. It was heartlessly disposed of in a mass grave along with hundreds of other Filipino patriots and innocent civilians whose lives were tragically ended by the ruthless Japanese Kempeitai. He did not receive the proper burial he so richly deserved.

Teodoro "Doring" Saavedra was one of the greatest fighters of the Philippines and a true icon of Cebuano eskrima. It is from Doring that my father learned the majority of his eskrima. Doring's hard-hitting style of eskrima is the basis of my Atillo Balintawak Eskrima to this day. His influence on me is undeniable, and I have made every effort to ensure his style of eskrima and memory are passed on to future generations.

Guerilla Recognition and the Koga Papers

From the General Headquarters (GHQ) of the South West Pacific Area (SWPA) Command in Melbourne, Australia, General MacArthur continued fighting the war in the Pacific and planning his strategic return to liberate the Philippines. He continued his support of Captain Cushing and the Cebuano guerillas. By December 1943, the guerilla movement in Cebu had grown significantly in strength. It was becoming increasingly more effective against the Japanese.

Recognizing the key to victory was dependent on the success of guerilla operations in the Philippines, General MacArthur officially recognized the Cebu Area Command (CAC) and guerilla forces throughout the Philippine Islands on February 12, 1944. He then retroactively promoted Captain James Cushing to the rank of Lieutenant Colonel and appointed him as the Commanding Officer of the CAC, 8th Military District of the U.S. Armed Forces in the Far East (USAFFE). Cushing restructured the CAC and transferred my dad and Delfin Lopez from the Combat Company of the 85th Infantry Regiment under Lt. Rogaciano "Popoy" C. Espiritu to "A" Company, 1st Battalion of the 87th Infantry Regiment under Lt. Col. Abel Trazo. MacArthur then began sending US Navy submarines loaded with weapons, ammunition, medicine, long-range radio equipment, and supplies from Australia to the guerillas in the Philippine Islands. These critically needed supplies were intended to support guerilla efforts and prepare for a large-scale campaign to recapture the islands from the Empire of Japan. The strengthened support by General MacArthur and supplies enabled Lt. Col. Cushing to broaden his efforts and improve intelligence coverage throughout Cebu and the neighboring islands. Also, Cushing was able to establish direct radio contact with General MacArthur in March 1944. This enabled Cushing to coordinate with MacArthur's headquarters in Australia and the other guerilla commanders throughout the Philippine Islands.

Lt. Col. James M. Cushing (b. 1910 – d. August 26, 1963) (U.S. National Archives and Records Administration).

A short time later, on the night of March 31, 1944, Admiral Mineichi Koga[54], the Commander in Chief of the Imperial Japanese Navy's Combined Fleet, and Rear Admiral Shigeru Fukudome[55], Koga's Chief of Staff, along with several staff officers, boarded two Japanese seaplanes and departed Korkor, Palau to Davao City on the Island of Mindanao, Philippines. Koga was afraid that the American naval forces were closing in on his headquarters on the island of Palau. Admiral Fukudome boarded the plane with a leather pouch that contained Combined Fleet Secret Operations Order No. 73. Known as the "Z Plan," the secret military plan outlined the Japanese military's strategy to protect areas in the South Pacific under their control and

---

[54] Admiral Mineichi Koga (September 25, 1885 - March 31, 1944) was promoted to Marshal Admiral posthumously after the war and given a full state funeral in Tokyo, Japan.

[55] Shigeru Fukudome (February 1, 1891 - February 6, 1971) was Chief of Staff of the Imperial Japanese Navy during World War II.

engage the American naval fleet in one final decisive battle in the Pacific. If successful, the American naval fleet in the Pacific would be crippled, and the Empire of Japan would have gained absolute military control of the entire Pacific Theater.

Rear Admiral Shigeru Fukudome (left) and Admiral Mineichi Koga (right). (U.S. National Archives and Records Administration).

Along the way, the two planes were met by an intense tropical storm. The storm caused the plane carrying Admiral Koga to crash into the Philippine Sea east of Mindanao, killing everyone on board. At approximately 2:30 a.m. on April 1, 1944, the second plane, carrying Admiral Fukudome, attempted to change course and land in Cebu City but crashed in the Bohol Strait off the coast of Cebu near Barrio Balud.

Fukudome and the survivors of the crash immediately began swimming away from the burning wreckage toward the coastline of Cebu. As the survivors approached the shoreline, they were quickly discovered by Ricardo Bolo, a Lieutenant of the Volunteer Guards (VG), his younger brother Edilberto, and their neighbor Valeriano Paradero. The Filipinos assisted the Japanese survivors to shore, where a large group of villagers had gathered along the beach at Magtalisay to investigate and aid in the rescue. Many of them were members of the VG. Realizing that the Japanese military would soon be looking for the survivors of the crash, Lieutenant Bolo and the members of the VG took the survivors prisoner. They immediately led them inland away from the visible coastline where they had come ashore.

A short time later, Lieutenant Bolo turned the prisoners over to Teopisto Tangub, the VG Commander of Barrio Sangat. Tangub then decided to transport the prisoners to Lt. Col. Cushing at the CAC headquarters at Tabunan. Not wasting time, Tangub and members of the VG assumed control of the prisoners and began making their way to Tabunan through the mountainous interior.

Recovery of the Z Plan Documents

Later that morning, a local fisherman named Pedro Gantuangko awoke from his sleep and was told about the crash by his neighbors. Gantuangko heard the roar of the seaplane engines earlier that night and the commotion on the beach throughout the morning. However, he elected to remain in bed until the commotion on the beach subsided. As he looked over the beach from his oceanside home, he observed something floating in the sea amongst the wreckage and debris from the crash. Gantuangko pointed the object out to his neighbor Rufo "Opoy" Wamer, who waded out and retrieved a wooden box covered in oil. Fearing they would be caught by the nearby Japanese soldiers who were combing the beach for survivors, Gantuangko and Opoy quickly transported the box to Opoy's residence. Once safely inside the house, they opened the oil-covered container. Inside they discovered a small pouch containing what looked like a handful of gold nuggets and a leather pouch containing bundles of papers written in Japanese. After separating the pages and lying the pieces of paper out to dry, they recognized that the number of Japanese soldiers searching the beach increased significantly. Realizing they may be looking for the documents he and Opoy recovered from the wreckage, they hurriedly buried the pouch containing the gold. They then relocated the leather pouch and sea-soaked Japanese papers to Gantuangko's mother's residence in the mountains further away from the beach.

For the next several days, Japanese soldiers intensified their hunt. They began searching houses along the shoreline for the documents they feared had floated to shore. Fearing he would get caught with the papers, Gantuangko sent word to Corporal Norberto "Berting" Varga, a local member of the Cebuano guerillas. He advised Berting that he and Opoy may have recovered something valuable from the wreckage. Berting was aware of the crash and the captured Japanese prisoners being transported to Cushing's headquarters at Tabunan by VG Commander Teopisto Tangub. Berting agreed to meet Gantuangko at his mother's residence. There, Gantuangko showed him the leather pouch and the Japanese military documents they had discovered a few days earlier. Berting immediately realized that the materials may be of significant military importance and decided to deliver the documents to Lt. Col. Cushing. Berting took custody of the pouch and began the long trek to Cushing's CAC headquarters in the mountains at Tabunan.

Prisoner Arrival at Tabunan

On the morning of April 3, 1944, Tangub and the Japanese prisoners arrived at the Command Post of "A" Company, 1st Battalion of the 87th Infantry Regiment, the unit of my dad and Delfin Lopez. After a short rest, Tangub was assigned additional guerillas to provide security along the way to Cushing's headquarters and departed. A guerilla *runner* was sent ahead of the party to recon the trail and alert Cushing of the arriving prisoners.

By April 8, 1944, the *runner* arrived and advised Cushing that Tangub and several guerillas would soon be coming with several Japanese prisoners. All of which survived the seaplane crash just off the coast of Cebu. Cushing immediately sent a coded radio message to General MacArthur at the South West Pacific Area Command (SWPA) headquarters in Australia. Cushing advised MacArthur that the prisoners were on their way and requested instructions of what action to take. Cushing was aware of the intensified efforts of the Japanese military to locate the prisoners and added: "constant enemy pressure makes this situation very precarious."

A short time later, Tangub and the Japanese prisoners arrived at Cushing's headquarters. Recognizing that three of the prisoners were seriously injured and needed medical assistance, Cushing immediately admitted them to the camp infirmary to receive medical treatment. He introduced himself and informed the prisoners that he was the commander of the Cebuano guerillas. As long as they were with him, they would be safe. He further advised the prisoners that he was aware they were the survivors of the seaplane crash and attempted to get their names and military rank. One of the prisoners spoke a little English and reluctantly answered Cushing's questions but remained evasive. The Japanese soldier's reluctance led Cushing to believe that the Japanese prisoners were hiding something from him. Cushing continued and learned that one of the prisoners was a flag officer. However, he was misled to believe it was General Twani Furomei, the Commanding Officer of Land and Sea Forces in Macassar, Celebes. In fact, it was Admiral Shigeru Fukudome.

On the afternoon of April 9, 1944, Cushing sent another coded radio message to General Douglas MacArthur. Cushing advised MacArthur that the prisoners were from the crashed seaplane on April 1, 1944. He then listed the prisoners by name, including General Twani Furomei, who he still hadn't discovered was actually Admiral Fukudome.

Filipino guerillas during WWII (U.S. National Archives and Records Administration).

Unbeknownst to Cushing, Col. Seiichi Ohnishi, Commander of the Ohnishi Butai, a highly-trained and well-equipped regiment named after himself, had received word of the crash. Col. Seiichi Ohnishi was also told that the survivors, including Admiral Fukudome, had been taken prisoner by the Cebuano guerillas. Suspecting they would be taken directly to Cushing; Ohnishi initiated a massive drive into the interior of Cebu to locate and recover the prisoners from Cushing and the Cebuano guerillas.

Almost immediately, Cushing received word that several columns of Japanese soldiers of the Ohnishi Butai were seen marching into the mountainous interior of Cebu toward his headquarters at Tabunan. Cushing then ordered the Cebuano guerillas at Tabunan to prepare for an evacuation on short notice and be prepared to move the prisoners to a safer location. The next morning, as the guerillas were finalizing their preparations to evacuate, gunfire began to ring out from the guerilla lookouts posted around the camp.

Ohnishi's soldiers had closed in during the early morning, and Cushing had been caught by surprise.

Cushing immediately orders the guerillas to gather the Japanese prisoners and follow him down a hidden trail that separated his Tabunan headquarters at Tupas Ridge from the adjacent mountain of Kamungayan. As they reached the bottom of the ridge and began their ascent up the slope to Kamungayan, they were in full view of the Japanese soldiers who had reached the vacated Tabunan headquarters. Suddenly, Cushing and the guerillas began receiving gunfire from a heavy Japanese machinegun positioned across the ridge. Almost immediately, several members of the group had been hit by gunfire and killed. The guerrillas attempted to return fire, but the small arms they possessed were ineffective and incapable of reaching the range of the Japanese machinegun. Cushing ordered the guerillas to disperse and spread-out as they continued up the ridge to Kamungayan. Within minutes, a Japanese spotter plane arrived and began shooting at the guerillas each time they moved from one cover position to the next as they made their way up the ridge. Finally, the guerillas were able to make their way to a rendezvous point where they were temporarily safe under an overhang of a cliff. After regrouping, Cushing and the guerillas continued their escape. Eventually, they made it to Kamungayan, where they established a temporary camp.

Once his radio operator set up his radio equipment at the temporary camp, Cushing realized that General MacArthur had not responded to the messages he sent before evacuating the camp at Tabunan. Cushing then called a conference of his top guerilla officers to discuss their options. The guerilla officers advised Cushing that their position at Kamungayan was entirely surrounded by Japanese soldiers. They further reported that innocent civilians were being rounded up and taken as hostages in Cebu City and several outlining villages outside the city. They wondered who the "General" really was and why Col. Ohnishi and the Japanese soldiers were desperate to recover him. As Cushing conferred with his officers, word began filtering in from his field commanders. Several more columns of Ohnishi's soldiers were closing in on his position and executing innocent civilians and burning houses along the way. Cushing and his guerilla officers unanimously agreed there was no other choice but to negotiate the release of the prisoners in exchange for the lives of the innocent civilians. As a result, Cushing decided to send Col. Ohnishi a message proposing that he refrain from further attacks on his guerillas and the innocent civilians in exchange for the Japanese prisoners.

Cushing then sent two volunteers from his contingent of guerillas, Corporal Herminio Cerna, and Numeriano Teves, along with two of the Japanese prisoners across the gorge to deliver the translated message to Ohnishi's soldiers across the ravine at Tupas Ridge.

Filipino guerillas with captured Japanese soldiers during WWII (U.S. National Archives and Records Administration).

  A short time later, the four messengers returned with cigarettes, Japanese liquor, sent by the Japanese as a gesture of goodwill, and a handwritten written reply from Col. Ohnishi himself demanding the immediate release of the prisoners. Ohnishi promised that the "properties and lives of Cushing's men and the civilians would be spared." In an attempt to stall for time, Cushing sent a reply back to Ohnishi offering to release four of the prisoners immediately and the rest in four days at the gates of the perimeter fence surrounding Cebu City. Ohnishi's response was immediate and straightforward, "All or no one at all."

  Cushing then went back to the "General" to discuss the situation and asked for his assurance that the killing and pillaging of the civilians would stop if he agreed to Ohnishi's terms. The "General" agreed, at which time Cushing sent a final message to Ohnishi, advising him that all of the prisoners would be released at daylight the following morning. Still waiting on a reply from General MacArthur, Cushing instructed his radio operator to send another coded message to MacArthur. Cushing advised MacArthur that the Japanese prisoners were "too hot to hold," and due to the number of civilians the Japanese soldiers were killing, he had made terms for their exchange with Col. Ohnishi.

The next morning on April 10, 1944, Cushing cautiously watched through his binoculars as Lieutenant Pedro Villareal, carrying a small white flag, escorted the Japanese prisoners and a small group of guerillas down a dry riverbed to a clearing along the bare mountain slope facing Ohnishi's soldiers. Moments later, a group of Japanese soldiers cautiously approached to make the exchange. The soldiers bowed ceremoniously to the "General," shook Villareal's hand and took custody of the prisoners. The only Prisoner of War exchange on Cebu during the war was complete. Villareal and the other guerillas quickly returned to Cushing's position and gave him two handwritten notes. One was from Ohnishi thanking him for treating the prisoners "kindly" and the other from the "General," extending his best regards for Cushing and his men.

Throughout the entire incident, Cushing had not received a response to his messages or any orders from General MacArthur. Finally, on April 11, 1944, Cushing received a response from the SWPA headquarters in Brisbane, Australia. The message instructed him to transport the prisoners to a rendezvous point in Southern Bohol or Southern Negros. They would then be extracted from the Philippines by submarine and transported to the SWPA headquarters in Australia. Unbeknownst to the SWPA headquarters and General MacArthur, Cushing had already exchanged the prisoners to spare the lives of the innocent Filipino civilians. Finally, on April 13, 1944, the SWPA headquarters in Australia read Cushing's messages sent days earlier. Not grasping why Cushing released a flag officer of the Japanese military, General Richard K. Sutherland, General MacArthur's Chief of Staff, sent a message back to Cushing, stating his actions were "reprehensible" and lead him to doubt Cushing's "judgment and efficiency." Acting on behalf of General MacArthur, Sutherland summarily discharged Cushing of his duties as the Commanding Officer of the CAC, 8th Military District of the US Armed Forces in the Far East (USAFFE). Cushing was devastated, but his loyalty to the Filipino people and guerillas under his command was of significantly greater importance than his formal rank and position. Without hesitation, Cushing ignored Sutherland's order to have him discharged as the Commanding Officer of the CAC and continued his campaign to liberate the island of Cebu from the Japanese.

An investigation ensued at the SWPA headquarters in Australia. Fortunately for Cushing, Colonel Courtney Whitney[56], an attorney and Cushing's former boss in the mining industry before the war, argued Cushing's position and was able to get the disciplinary action dropped and his rank and position reinstated.

Cushing and the Cebuano guerillas still didn't know the true identity of the "General" or why Ohnisi so ruthlessly and relentlessly pursued his release. Cushing then remembered the leather pouch and documents delivered to him by Corporal Norberto "Berting" Varga shortly after the "General" arrived at the Tabunan HQ on April 8, 1944. Upon inspection of the items, Cushing discovered a red leather pouch with significant Japanese characters printed in gold on the cover and several sealed envelopes and documents. As he examined the documents, he realized he had Japanese military plans of extreme importance. Cushing then sent a long, coded message to General MacArthur advising he was in possession of several Japanese operations maps showing airbases, naval bases, wireless stations, emergency landing fields, triangulation points, heights, and other control symbols in the legends. He further advised he had field orders and additional maps of Palao, Philippines, French Indo-China, Hainan, and Southern China. He stated he would be sending the documents to Col. Andrews in Southern Negros where they could be transported by submarine to General MacArthur in Australia.

On April 15, 1944, Cushing packed the documents in several empty mortar shells and gave them to two former American prisoners of war, Russ Snell and Jimmy Dyer. Cushing then assigned a small contingent of experienced Cebuano guerillas as added security and instructed them to deliver the documents to Col. Andrews in Southern Negros. Thirteen days later, on April 28, 1944, the group arrived at Col. Andrews' headquarters in Southern Negros, where they turned the documents over to Andrews.

---

[56] Major General Courtney Whitney (May 20, 1897 - March 21, 1969) was a lawyer and senior official for the South West Pacific Area (SWPA) Command during the war under the direction of General Douglas MacArthur.

On May 11, 1944, the documents were turned over to Lt. Cmdr. Francis David Walker, Jr, Commanding Officer of the U.S. submarine, the USS Crevalle. Lt. Cmdr. Walker had been dispatched to the coast of Negros to retrieve the documents. The submarine quickly got underway and transported the materials to Australia, where they eventually made their way to the SWPA headquarters in Brisbane on May 21, 1944. The documents were quickly translated by the Allied Translation and Interpreter Section (ATIS) and delivered to General MacArthur. Allied commanders distributed the plans to other commanders who were able to exploit and use the secret Japanese military plans to develop a counter-offensive strategy in the Pacific. This counter-offensive ultimately led to the liberation of the Philippine Islands and the defeat of Japanese forces in the Pacific.

The recovery of the Z Plans and the sacrifices made by Lt. Col. Cushing and the Cebuano guerillas to safeguard the documents and facilitate the transport of the plans to General MacArthur in Australia is considered one of the most important military intelligence successes of World War II.

Death of Governor Hilario Abellana

Col. Ohnishi temporarily kept his word to stop killing innocent civilians and pursuing Cushing and the Cebuano guerillas. However, he still needed to recover the missing Z Plan documents from the seaplane crash on April 1, 1944. Unaware that the plans had been secretly transported to General MacArthur in Australia by submarine, Ohnishi continued to use Filipino undercovers to gather information and determine the whereabouts of the documents.

Out of frustration, Ohnishi turned his attention to Governor Hilario "Dodong" Abellana. A few years earlier, in 1942, Abellana had headed efforts to secretly assemble a group of patriots from Mambaling and organize one of the first guerilla units on Cebu. This secret unit included several eskrimadors, including my dad and Delfin Lopez, as well as Lorenzo "Tatay Ensong" Saavedra's nephews Doring and Frederico "Pedring" Saavedra. Abellana decided to remain the Governor of Cebu throughout the Japanese occupation of Cebu. He felt he could better support the resistance movement by staying close to the Japanese government and filter intelligence information to the guerillas. His role as governor of Cebu was becoming less significant to the Japanese, and on January 17. 1943 he escaped to the neighboring island of Bohol.

Ohnishi relentlessly searched for Abellana in hopes that he knew of the missing Japanese plans. The Japanese military had already arrested and tortured most of Governor Abellana's relatives to locate him. None were willing to give up his location. On July 13, 1944, out of desperation, Ohnishi ordered the arrest of Governor Abellana's cousin, Jovito Abellana. Jovito was arrested and brought to the headquarters of the dreaded Japanese Kempeitai at the Cebu Normal School, where they brutally tortured him for weeks. Eventually, the Kempeitai were convinced Jovito did not know of the whereabouts of his governor-cousin, and they imprisoned him. Throughout his imprisonment, Jovito witnessed the daily torture and execution of hundreds of Filipinos. Many were taken by the Kempeitai, who gathered prisoners from their cells twice a week and transported them to a location outside Cebu City. They were ruthlessly and callously beheaded in mass executions. Presumably, to make room at the Kempeitai headquarters for more captured and imprisoned Filipino prisoners.

The Japanese eventually captured governor Abellana on September 3, 1944. They took him to the Kempeitai headquarters at the Cebu Normal School, where he spent the remaining months of his life being beaten and tortured by the Kempeitai. On January 15, 1945, the Japanese executed him. His body was disposed of in the same manner as Teodoro "Doring" Saavedra and never recovered.

MacArthur's Return

Finally, on October 17, 1944, Allied forces led by General Douglas MacArthur executed Operation King Two.[57] It began with amphibious assaults on the islands of the Leyte Gulf. Three days later, on October 20, 1944, after hours of heavy naval gunfire, Allied forces landed on the sandy beaches of Leyte Island, east of Cebu, and began their advance inland. As the landing force cleared the beach, General MacArthur, accompanied by Philippine President Sergio Osmeña, walked ashore at Palo Beach, Leyte, Philippines, fulfilling his promise to return to the Philippines and liberate the islands from the Empire of Japan. By December 26, 1944, the Allied forces

---

[57] King Two was the code name given to the plan to conduct an amphibious invasion of the Gulf of Leyte and initiate the military campaign to recover the Philippines from the occupation forces of the Empire of Japan.

had completed the first phase of the operation and recaptured the island of Leyte.

General Douglas MacArthur wades ashore at Palo Beach, Leyte, Philippines accompanied by Philippine President Sergio Osmeña, October 20, 1944. (U.S. National Archives and Records Administration).

The Allied forces at Leyte immediately began planning and preparing for the second phase of the operation, including amphibious landings on the islands of Cebu, Bohol, and Negros. Cebu was still under the control of approximately 14,500 Japanese soldiers, of which 2,000 were contained in the northern area of Cebu by the 8,500 Cebuano guerillas under the command of Lt. Col. James Cushing. Japanese forces in Cebu were ready with an extensive network of formidable defensive positions around Cebu City.

Liberation of Cebu

On the morning of March 26, 1945, Major General William H. Harold launched Operation Victor II[58]. After an hour-long naval bombardment of Cebu City and the coastal areas of the island, Allied forces consisting of the

---

[58] Operation Victor II was the code name given to the military plan to assault and recover the islands of Cebu, Bohol and the southeastern area of Negros.

Cebuano guerillas march through the streets of Cebu City, Cebu, Philippines, April 6, 1945. (U.S. National Archives and Records Administration).

Americal Division's 132nd and 182nd Infantry Regiments landed at Talisay Beach just a few miles south of my home in Mambaling. The Allied forces encountered very little resistance from the Japanese during the initial assault. However, they suffered heavy losses from mines and booby traps left behind by the Japanese as they came ashore and negotiated the beachheads along the coast. Engineers quickly cleared lanes through the minefields so additional forces could safely come ashore. Once on shore, the Americal Division slowly made its way inland toward Cebu City, meeting only small pockets of Japanese resistance. Allied forces then advanced into Cebu City the very next day. They were able to secure the harbor and capture Lahug Airfield and nearby Mactan Island by March 28, 1945.

The remaining Japanese forces remained resilient and established strong defensive positions along the mountains and areas north of the city. Heavy fighting between Allied forces and the Japanese began as the Japanese grew increasingly more determined to maintain control of Cebu and hold their positions along the outer edge of the city. With the help of the Cebuano guerillas, Allied Forces were able to capture the barangay of Pari-an on March 29, 1945, and liberate the barangay of T. Padilla by April 7, 1945. As Allied

Tanks of the 716th Tank Battalion transport troops of the Americal Division during clean-up operations in Northern Cebu, Philippines, May 8, 1945. (U.S. National Archives and Records Administration).

forces and the Cebuano guerillas continued their offensive, Japanese forces withdrew even further into the mountains and scattered seaports along the coast. Barrages of naval bombardments forced the Japanese along the coast to withdraw inland toward the area of Fort San Pedro in Cebu City, where they were trapped. Final mop-up operations were conducted along the coast, forcing many Japanese soldiers to surrender. However, several soldiers in the area made one final, last stand near the border of Mandaue City[59]. Allied forces continued their offensive and were able to secure the remaining areas of Cebu City on April 8, 1945.

For the next several days, unrelenting naval bombardments continued in the mountainous areas surrounding Cebu City. Allied forces and the Cebuano guerillas pressed forward, conducting clearing operations. Several battles were fought against the enduring Japanese troops who had heavily

---

[59] Mandaue City is located on the central eastern coast of Cebu and is one of three highly urbanized cities on Cebu that forms the Cebu Metropolitan area.

entrenched themselves along the outskirts of the city. Within three days, Allied forces had secured the area and retaken Gochan Hill north of Cebu City.

On April 13, 1945, Allied forces and the Cebuano guerillas, including my dad and Delfin Lopez, led a two-pronged division level attack against the Japanese. The attack successfully forced the remaining Japanese troops into the northern mountains of the island. Within a week, Allied forces and the Cebuano guerillas followed up their initial attack. They began conducting harassment operations against the remaining Japanese soldiers in the northern mountains to keep them from regrouping and mounting an offensive counter assault. The Americal Division ceased operations on June 20, 1945, and withdrew to Cebu City to prepare for future military operations. The rest of The Cebuano guerillas under Lt. Col. James Cushing, including my dad and Delfin Lopez, continued fighting until the final Japanese stronghold in the barangay of Minglanilla was captured on July 2, 1945.

Japanese Surrender

On July 26, 1945, Allied Forces issued the Potsdam Declaration[60] calling for the immediate, unconditional surrender of Japan. The declaration and ultimatum stated that if Japan did not surrender, it would face "prompt and utter destruction." Japan did not respond to the warning and essentially rejected the conditions of the surrender. In a subsequent effort to bring an immediate end to the war and prevent a massive Allied assault on Japan, the United States dropped an atomic bomb on the Japanese city of Hiroshima on August 6, 1945. This was followed by a second atomic bomb on the city of Nagasaki on August 9, 1945.

Six days later, after grasping the massive destruction caused by the atomic bombs and the likelihood of a large-scale Allied invasion of their homeland, The Empire of Japan announced its surrender on August 15, 1945.

Almost immediately, the Cebu Area Command (CAC) decommissioned many guerila units that encompassed the island command. Subsequently, my dad was honorably discharged from "A" Company, 1st Battalion of the 87th Infantry of the United States Armed Forces Far East

---

[60] The Potsdam Declaration is a declaration issued by US President Harry S. Truman, US Prime Minister Winston Churchill and Chairman of the Nationalist Government of China, Chiang Kai-shek calling for the surrender of all Japanese armed forces during World War II.

(USAFFE) on August 15, 1942, along with Delfin Lopez. Both eskrimadors served honorably throughout the war alongside their lifelong childhood friend and compatriot, Teodoro "Doring" Saavedra, who tragically lost his life a few years earlier.

On August 19, 1945, Lt. Gen. Kataoka Tadasu, Commanding General of the 1st Division of the Japanese Imperial Army, arrived in Cebu City to begin surrender negotiations. The surrender of the Japanese military on the island was one of the first official large-scale surrenders to occur in the Pacific. 8,500 Filipino guerillas and 14,900 US troops fought against 15,000 Japanese soldiers to liberate the island of Cebu. In the end, approximately 410 American and Filipino personnel died, and roughly 1,700 were wounded. The Japanese faired much worse, however. The Japanese military suffered over 9,000 casualties, and approximately 2,000 committed *seppuku*, a form of Japanese ritual suicide.

Surrender of the Empire of Japan to General Douglas MacArthur onboard the USS Missouri, September 2, 1945. (U.S. National Archives and Records Administration).

On August 28, 1945, Lt. Gen. Kataoka Tadasu, along with thousands of Japanese soldiers, marched out of the densely forested hills of Cebu into an open field near Ilihan, Cebu, Philippines. Tadasu surrendered his sword to Maj. General William H. Arnold of the 23rd Infantry Division[61] formally surrendering the island of Cebu and all Japanese forces on the island to Allied

---

[61] The United States Army 23rd Infantry Division, also known as the Americal Division, was activated 27 May 1942 on the island of New Caledonia. The name is a contraction of "American, New Caledonian Division".

Forces. Two weeks later, on September 2, 1945, the Empire of Japan formally surrendered to General Douglas MacArthur with the signing of the Japanese Instrument of Surrender on board the USS Missouri in Tokyo Bay, officially ending the war.

Crowds of people swarm around a street in Cebu City, Cebu, Philippines, to watch a parade of the 77th Infantry Division during the celebration of V-J Day, September 7, 1945. (U.S. National Archives and Records Administration).

Passing of an Eskrima Legend

Two months after the Japanese surrendered, Lorenzo "Tatay Ensong" Saavedra died peacefully at his home in Mambaling of natural causes at the age of 93. His life was one of challenge and adversity. As a child under the oppression of Spanish colonialism, he stood up against subjugation and abuse. He was incarcerated in the Cebu Provincial Jail by the Spanish authorities for over half his life for insurrection[62]. He had lived through the latter part of the

---

[62] An open resistance against the orders of an established authority.

Spanish Colonial Era[63], the Spanish-American War[64], the Philippine-American War[65], and World War II. He had witnessed firsthand the evolution of his native homeland from an oppressed nation under foreign rule to an independent and free sovereign democracy. It was only after his release from the Cebu Provincial Jail in 1920 at the age of 68 that he was able to willingly share his unique method of eskrima and enjoy his newfound freedom and peace. Unfortunately, that peace was short-lived with the outbreak of World War II and the untimely murder of his beloved nephew and protégé, Teodoro "Doring" Saavedra, at the hands of the Japanese Kempeitai.

Nevertheless, Tatay Ensong lived a life of a legend and was the single most influential eskrimador in the history of Cebuano eskrima. If not for his innovations while imprisoned, and his unyielding desire to share his art with those eager to learn, many styles of eskrima practiced around the world today would not exist. This is certainly true for the art of Balintawak eskrima, and the various styles of Balintawak taught today.

My dad, Delfin Lopez, and many of Tatay Ensong's remaining family members who survived the war gathered together and buried him in the front yard of Delfin Lopez's house a short distance from my home in Mambaling. His remains stayed there until years later when the family of Delfin Lopez sold the house, and an apartment building was erected in its place.

I was very fortunate to have had the privilege of knowing Tatay Ensong when I was a young boy. He and I lived in the same neighborhood in the Mambaling District of Cebu City. My family was very close to the Saavedras, and my father was a childhood friend of Tatay Ensong's nephew, Teodoro "Doring" Saavedra. One of my earliest memories was of Tatay Ensong and Doring practicing eskrima together. In one instance, I recall watching them train with a spear. Tatay Ensong would throw the spear at

---

[63] The Spanish Colonial Era (March 16, 1521 – December 10, 1898) of the Philippines began with the arrival of Ferdinand Magellan in 1521 and ended with the signing of the Treaty of Paris in 1898 relinquishing the Philippines to the United States.

[64] The Spanish-American War (April 21, 1898 – August 13, 1898) was fought between the United States and Spain and ended with the signing of the Protocol of Peace in Washington, D.C.

[65] The Philippine-American War (February 4, 1899 – July 2, 1902) was fought between the United States and the newly acquired Philippines subsequent to Treaty of Paris and ended with the signing of the Philippine Organic Act on July 1, 1902.

Doring as he stood in front of a tree. Doring would then parry it to the side as it passed by so it wouldn't strike him. I was also able to watch practice sessions with Doring and my dad, as well as training sessions with Venancio "Anciong" Bacon and Delfin Lopez. The four eskrimadors were childhood friends and would gather at our home in Mambaling nearly every week to practice. Even as a child, I observed that Doring's style of eskrima was hard-hitting and aggressive. He was a significantly better eskrimador than the others except for Anciong. As a student of Tatay Ensong and childhood friend of Doring's, Anciong was very talented and was the best fighter at the time, second to only Doring.

The Aftermath of War

In the aftermath of the war, my dad and Delfin Lopez had changed. My dad didn't talk much of the war. He immediately went back to work as a supervisor for the Health Rubber Company, a shoe manufacturing company in Mambaling, which had previously been run by the Japanese before the war. Delfin Lopez joined the Cebu City Police and became a member of the Cebu City Police Department Secret Service Force.

Remains of the Balintawak Beer Brewer Company, Inc. following the war. (U.S. National Archives and Records Administration).

As proud Filipino patriots, they were some of the first Filipinos to join the fledgling guerilla resistance movement led by Lieutenant José Macabuhay and governor Hilario Abellana. As eskrimadors, my dad, Doring Saavedra, and Delfin Lopez had seen the realities of war and used the art of eskrima in actual life and death combat. All of the Casualty and

Wounded in Action Reports written by Lt. Col. Cushing during the war reflected that most guerilla engagements were in close quarters in the jungles and densely forested mountains of Cebu. The majority of deaths occurred by bayonet thrusts, close pistol engagements, and the use of the bolo to kill the enemy.

Declassified Casualty Report drafted by Lt. Co. James Cushing listing Vicente Atillo (#114) as being killed in action March 27, 1945. (U.S. National Archives and Records Administration).

My dad was wounded on several occasions and was even identified as being killed in action (KIA) on March 27, 1945, in a Casualty Report[66] drafted by Lt. Col. Cushing. As a result of their experiences in combat, the style of eskrima practiced by the elder Teodoro "Doring" Saavedra and taught to my father and Delfin Lopez had evolved. It was no longer the eskrima taught to them by Lorenzo "Tatay Ensong" Saavedra during their childhood or as members of the Doce Pares Club before the war. Out of necessity for survival, it had evolved into a modern variation of the art that emphasized the application of eskrima in modern close-quarters combat.

For my dad's service as a member of the famed "A" Company, 1st Battalion of the 87th Infantry of the United States Armed Forces Far East

---

[66] US National Archives and Records Administration, 2-1 Cebu Area Command, From RG: 407, Guerrilla Unit Recognition Files National Archives Identifier: 1430747 Container Identifier: 270 HMS Entry Number(s): A1 1095, Creator: Southwest Pacific Area. 4/18/1942-9/2/1945.

(USAFFE) under Lt. Col. Cushing, he was awarded several medals and citations for his bravery and actions in combat. They included the Asiatic Pacific Campaign Medal (3 Bronze Stars), WWII Victory Medal (3 Bronze Stars), and Philippine Liberation Ribbon (2 Bronze Stars). Additionally, on December 14, 2016, the U.S. Congress passed Public Law 114–265, awarding the Congressional Gold Medal[67] to all Filipino veterans and guerillas who served during World War II for their dedication to service.

4 Former Japanese Officers Are Hanged

Manila—(P)—Four former Japanese officers, convicted of torture slayings in the Philippines, have been hanged, headquarters of army forces, western Pacific, announced today.

The four were: Tsunau Toyagana, former military police lieutenant colonel convicted of responsibility in the deaths of 500 Filipinos and Americans at Ft. Santiago; Ens. Kagajiro Mukai, head of a naval raiding party responsible for the murder of 50 Filipinos at Davao; Lt. Col. Isamu Morimoto and Capt. Kasatoro Tsuruyama, jointly convicted of slaying two American prisoners at Cebu City.

Daily Globe Newspaper story describing the hanging of Capt. Kasatoro Tsuruyama in Tokyo, Japan, subsequent to a War Crimes Tribunal, June 6, 1946, Manila, Philippines.

My dad, Teodoro "Doring" Saavedra, Delfin Lopez, and thousands of other Filipinos participated in the war out of national pride and a commitment to the struggle against oppression. In exchange for their sacrifice and service in the United States Armed Forces of the Far East (USAFFE), U.S. President Franklin D. Roosevelt issued the Second War Powers Act[68] (Public Law 77–507; 56 Stat. 176) on February 1942. The Act promised a simplified naturalization process and American citizenship for Filipinos who served in the United States Armed Forces as well as full veteran's benefits. The Act included Filipinos who served in the Commonwealth Army of the Philippines, Philippine Scouts, and later members of the Recognized Guerillas. Unfortunately, after the election of US President Harry S. Truman on April 12, 1945, and the

---

[67] 114th Congress, Public Law 114-265, Filipino Veterans of World War II Congressional Gold Medal Act of 2015, enacted on December 14, 2016, to award a Congressional Gold Medal, collectively, to the Filipino veterans of World War II, in recognition of the dedicated service of the veterans during World War II.

[68] The Second War Powers Act was passed on March 27, 1942. The act further strengthened the executive branch powers towards executing World War II. This act allowed the acquisition, under condemnation if necessary, of land for military or naval purposes. Provisions of the act also reduced the naturalization standards for aliens within the U.S. Armed Forces.

surrender of Japan on September 2, 1945, President Truman and the US Congress rescinded the citizenship and veterans benefits provided by the Second War Powers Act by passing the Rescission Act of 1946 (Public Laws 79–301 and 79–391; 60 Stat. 6 and 60 Stat. 221)[69] on February 18, 1946. The Rescission Act of 1946 did not consider the wartime service of the Filipinos to be active military service and, therefore, did not qualify for benefits.

This decision was a shock to all Filipinos who proudly served as members of USAFFE, including my dad. Instead of receiving the veteran's benefits and American citizenship initially promised to him in exchange for his military service and assured by the Second War Powers Act at the time of his induction into the USAFFE, my dad received a meager retirement pension of 500 pesos. Even by today's standard, 500 pesos is valued at just over $9.00.

In the end, over 57,000 Filipinos died in service to USAFFE, and an estimated one million Filipino civilians were casualties of the war. Only a small number of Filipinos who survived the war were able to obtain U.S. citizenship before the Rescission Acts of 1946. The Filipino veterans of WWII, many of them eskrimadors, fought, suffered, and in many instances, died in the same manner and under the same commander as other members of the United States Armed Forces during WWII. These Filipinos were not only fighting for their homeland. They were also defending, and ultimately liberating, sovereign territory held by the United States Government.

---

[69] The Rescission Act of 1946 is a law of the United States that retroactively annulled benefits that would have been payable to Filipino troops for their military service under the auspices of the United States during the time that the Philippines was a U.S. territory.

CHAPTER 5
Balintawak

Soon after the WWII ended, many of the original members of the Doce Pares Club reunited and resumed training. However, with the deaths of Lorenzo "Tatay Ensong" Saavedra and Teodoro "Doring" Saavedra, the club was now under the leadership of Eulogio "Yoling" Cañete and his younger brothers Florentino and Filemon "Momoy" Cañete. Although the Cañetes had been students of the elder Saavedra before the war, they possessed a different perspective of how the art of eskrima should be taught. That perspective significantly impacted the direction of the club and conflicted with many students loyal to the Saavedras and the Saavedra style of eskrima.

Internal Bickering and Politics

Not long after the club was reestablished, jealousies and petty rivalries erupted between several instructors and their respective students. Particularly between Venancio "Anciong" Bacon and Ciriaco "Cacoy" Cañete. The internal disagreements and constant infighting were not exclusive to the rival Saavedra style and Cañete students. Differences began to erupt between the older Cañetes and the younger family members who were now *magtutudlo*[70] within the club. The elder members of the family, like Filemon "Momoy" Cañete, preferred to teach the old style of eskrima taught by Tatay Ensong before the war. In contrast, many of the younger members, like Cacoy, chose to modify the old style and add techniques from the Japanese martial arts of *karate*, *ju-jitsu*, and *judo* and implement *katas*[71] and a Japanese style grading system using colored belts.

This perspective did not sit well with many of the Doce Pares Club

---

[70] *Magtutudlo* is a Cebuano term that refers to an instructor or teacher. It is a more accurate term than the more commonly used term, *guro*, which is a term used to refer to a teacher in an academic setting.

[71] *Kata* is a Japanese term that describes a series of choreographed techniques and movements common to Japanese martial arts.

old-timers like Venancio "Anciong" Bacon, Delfin Lopez, and my dad, Vicente "Inting" Atillo. They were close friends and students of Doring. Lopez and my dad fought alongside Doring during the war as members of the famed Cebuano guerilla forces under Lt. Col. James Cushing and Doring had been brutally tortured and executed by the Japanese Kempeitai. The acceptance of Japanese customs and the implementation of Japanese *budo* practices into the closely guarded Filipino art was unacceptable and disrespectful to the memory of those who fought against the Japanese during the war. Additionally, the experiences shared by many eskrimadors during the war had changed the way they viewed the art and its application to modern warfare. This included a shift in focus from the more traditional long-range techniques adapted from the Spanish to the close-range blade and stick methods more appropriate for close-quarters combat.

Two very different factions began to form and further divide the remaining members of the post-war Doce Pares Club. Students and instructors began to establish separate loyalties. They split into two very distinct groups - those loyal to the Saavedras and those devoted to the Cañetes.

Venancio "Anciong" Bacon

Venancio "Anciong" Bacon was born on October 15, 1912, in Carcar, south of Cebu City. Shortly after his birth, his family moved to the *barangay* of San Nicolas in Cebu City, where Anciong became a childhood friend of Teodoro "Doring" Saavedra. As children, the two youngsters learned eskrima from Tatay Ensong, who groomed them into talented eskrimadors.

Venancio "Anciong" Bacon, c. 1952.

For several years following the re-establishment of the Doce Pares Club after the war, Anciong often discussed his increasing disgust for the direction of the club and his growing dislike for the Cañetes with my dad and Delfin Lopez. In particular, his hatred for Cacoy, who had become one of the chief instructors of the post-war club. Anciong felt that the Cañetes were no longer practicing true combative eskrima and focused on ineffective and old-fashioned techniques. In addition, he believed the Cañetes were using the Doce Pares Club for financial gain and commercializing and diluting the Filipino art by incorporating Japanese martial arts and techniques from *judo* and *jiu-jitsu*. Anciong's loyalty had always been to the Saavedra family. He had been a faithful student of Tatay Ensong since childhood and was a close friend and training partner of Doring, who was known to be the greatest fighter of his time. Standing at only 5 ft. 2 inches, Anciong was a tiny man but was a master tactician and feared eskrimador. He had also practiced *dumog*,[72] *Combat Judo*,[73] and *boxing*.[74] His years of training with Doring and his experience as a fighter gave Anciong a sense of what works in actual combat. He was passionate about preserving the art taught to him by Tatay Ensong and the combative nature of Filipino eskrima.

Delfin Lopez, c. 1952.

I was present during many of these discussions and remembered how frustrated Anciong had become. Delfin and my dad were often successful in reassuring Anciong and convincing him to remain a member of the Doce Pares Club to continue teaching the Saavedra style. However, by December 1951, Anciong finally had enough and decided to disassociate himself entirely

---

[72] A Filipino style of wrestling and is a term that is commonly used to refer to the grappling aspect of the Filipino martial arts.

[73] A term used to describe empty hand self-defense techniques popular throughout the post-war Philippines. The term was adopted during World War II by American service members stationed in the Philippines.

[74] American-style Boxing was brought to the Philippines by American soldiers after the Spanish-American War. Boxing was legalized in the Philippines in 1921 and has become one of the most popular sports in the Philippines, second to only basketball.

from the Cañetes and establish a new club focused on the continued propagation of the Saavedra style and the combat applications of Filipino eskrima. Anciong was also a destitute man and wanted to teach eskrima to make money. However, he had little interest in commercializing the art.

I was at home in Mambaling with my dad and Delfin Lopez when Anciong made up his mind and announced he was finally leaving the Doce Pares Club. He said he wanted to establish a separate club with a new name and identity. Lopez and my dad had also grown tired of the internal politics and growing rivalries within the Doce Pares Club and were ready to leave as well.

Vicente "Inting" Atillo, c. 1952.

Although Lopez and my dad supported the idea of splitting from the Cañetes and creating a new club, they disagreed with Anciong's decision not to use the *Doce Pares* name. The original Doce Pares Club had been founded by Tatay Ensong in 1932. The name was chosen by Tatay Ensong to honor the Frenchman he befriended while incarcerated in the Cebu Provincial Jail as well as symbolically represent the Twelve Peers of France. Furthermore, the *Doce Pares* name was now famous throughout the Philippines because of the legendary exploits and reputation of Doring. My dad recommended they keep the name but modify it to reflect the Saavedra style and pay tribute to Tatay Ensong and Doring. As an alternative, he suggested they call the club the *Doce Pares Club of the Saavedras* or the *Saavedra Doce Pares Club*. However, after hours of heated debate, Anciong maintained his position and insisted the name be changed. Anciong held that a new name would be chosen that would completely disassociate the new club from the post-war Doce Pares Club and the Cañete family.

The decision to abandon the name infuriated Lopez, who immediately began to argue with Anciong. Although the two were close friends and training partners, there had always been tension between the two eskrimadors. Both felt they were the best fighters next to Doring and often quarreled over who was the better eskrimador. Lopez liked the idea of starting a new club but felt that Anciong's decision to stop using the *Doce Pares* name was motivated by money. Lopez further alleged that by electing himself as the chief instructor, Anciong distinguished himself as the top fighter and better eskrimador between the two. Anciong became more and more irritated as Lopez continued arguing, and his temper began to escalate. He remained firm in his decision to rename the club and his choice to be the chief instructor. Anciong's defiance angered Lopez even more, at which time he jumped up from his chair and challenged Anciong to fight. Right away, Anciong accepted and squared off with Lopez. Recognizing the situation was rapidly getting out of control, my dad quickly stepped between the eskrimadors to prevent his two friends from fighting. He then told Lopez that he could not beat Anciong and pleaded with them not to fight. This further enraged Lopez, who continued his challenge. My dad was eventually able to calm them down. Still, Anciong and Lopez left our house in Mambaling without a compromise being reached between the two eskrimadors.

Venancio Bacon vs. Delfin Lopez

The dispute between Anciong and Lopez continued for weeks following the argument. My dad was well-liked by all eskrimadors and was known as a peacekeeper who would often be the voice of reason between quarreling rivals. He persistently tried to ease the tension between the two famed eskrimadors yet was unsuccessful. Finally, Anciong and Lopez agreed to resolve their disagreement by fighting in Tina'an, a small town in southern Cebu. I was there with my dad, as were several other students and friends interested in watching the two eskrimadors fight. Anciong and Lopez were both well-known and celebrated eskrimadors, and the idea of them fighting each other excited everyone. Challenge matches occurred often and were a common means of resolving disputes between rival eskrimadors. An eskrimadors pride and reputation were on the line, and matches were taken seriously.

Before the match, my dad again pleaded with Anciong and Lopez. He assured them a compromise could be reached without friends coming to blows. My dad told Lopez that he could not beat Anciong, and as the founding *grandmaster*[75] of the club, he had the right to change the name to whatever he wanted. Disregarding my dad's plea and attempt to prevent them from fighting, Anciong and Lopez elected to continue.

Vicente "Inting" Atillo practicing with a young Crispulo "Ising" Atillo shortly after the establishment of the Balintawak Self Defense Club, c. 1954.

As the fight began, Lopez rushed forward and began striking Anciong repeatedly in the legs. Unaffected by the strikes, Anciong promptly countered with a series of rapid counter-strikes that quickly overwhelmed Lopez. Unable to defend himself against Anciong's onslaught of blows, Lopez grudgingly conceded to Anciong and surrendered. As with most challenge-matches between talented eskrimadors, the fight ended as quickly as it began. Reluctantly, Lopez agreed to accept Anciong's decision to stop using the *Doce Pares* name and decided to support his position as the chief instructor

---

[75] The term *Grandmaster* is a title commonly used to describe the head master or head instructor of a martial arts style or organization. It was not used in the Filipino martial arts until the late 1960s or early 1970s. Instead, the title of Head Instructor or Chief Instructor was more commonly used.

of the forthcoming club.

As the spectators dispersed and the two eskrimadors settled their personal differences, my dad and I talked about the fight as we returned home to Mambaling. My dad told me that Anciong was the best fighter in Cebu and was second to only Doring. He said that Anciong's style was different, and nobody could beat him. My dad further advised that Anciong was very poor and uneducated and needed to start his own club to make money and provide for his family. He reminded me that Anciong was the godson of my grandfather. As a family, we would support him and become members of his club.

The Balintawak Self Defense Club

On June 28, 1952, Anciong organized a friendly gathering and practice session at a nearby outdoor recreation area close to my home in Mambaling. Anciong was a very well-known member of the Doce Pares Club and had many devoted students. Most of whom were dedicated practitioners of the Saavedra style and experienced eskrimadors who were experts in their own right, having learned other methods of eskrima before joining the club. I was also there, along with several other younger students who were mostly family members of the eskrimadors participating in the training.

Even though I had been learning eskrima from my dad and Delfin Lopez from when I was a child, I didn't start practicing in earnest until 1952, when I was 14 years old. Anciong would often practice with me and ask that I demonstrate at local fiestas. He would smile and chuckle at my tenacity and fearlessness as I tried to best the older students and prove myself as a young eskrimador.

After several hours of training, Anciong gathered everyone together to eat *lechón*[76] and mingle. Anciong then asked for everybody's attention and formally announced he was leaving the Doce Pares Club. Anciong explained that he felt the Cañetes were no longer practicing true combative eskrima. He stated that the Cañetes were diluting and commercializing the Filipino art by incorporating Japanese martial arts and adding techniques from *judo* and *jiujitsu*. He further noted that the Cañetes were using the Doce Pares Club for

---

[76] *Lechón* is a Spanish pork dish popular throughout the Philippines that features a whole roasted pig slow roasted over charcoal.

financial gain and to advance their political ambitions. As a student of Tatay Ensong and a childhood friend of Doring, Anciong felt he needed to disassociate himself completely from the Cañetes and form a new club that focused entirely on the combat applications of Filipino eskrima.

Anciong's students were excited and welcomed the idea of having their own club. They had also grown tired of the internal politics and rivalries and were ready to leave the Doce Pares Club as well. The group began discussing the specifics of the new club and a suitable location where members could get together and practice. Anciong suggested that because he, Lopez, and my dad all lived close to each other in the same *barangay*, the club's headquarters should be in Mambaling. Some of the students were receptive to the idea. However, many expressed concerns over traveling to Mambaling from Cebu City, where most of them lived and worked. Eduardo Baculi, a watch repairman and student of Anciong, interjected and generously offered the use of a small space adjacent to a pigsty behind his watch repair shop located on Balintawak Street in the Colon District of Cebu City.

The first elected President of the Balintawak Self Defense Club, Atty. Eulalio E. "Jose" Causing. (Photo courtesy of David Ducay and Livinia Causin Ducay).

The majority of the group liked the idea. After a few minutes of discussing potential locations, Anciong accepted Baculi's offer and announced that the club would meet behind Baculi's shop on Balintawak Street. The discussion then focused on what the club should be named. Anciong advised that since the club would be headquartered on Balintawak Street behind Baculi's shop, it would be called the Balintawak Self Defense Club.

The name served two purposes. First, it identified the location of the club, and second, it symbolized Anciong's struggle against the Doce Pares Club and his desire to propagate the Saavedra style. To Filipinos, the word *balintawak* is commonly associated with a war cry that signified the beginning of the Philippine Revolution against Spain in August 1896.

Once the name and practice location was decided upon, an election of officers was conducted by those in attendance. It was agreed that Anciong would serve as the Chief Instructor of the club, Atty. Eulalio E. "José" Causing would be the first elected president, and Atty. Teodoro Arnoco and Atty. Cecilio de la Victoria would be the officers of the club. Other original members included Delfin Lopez, the great Timoteo "Timor" Maranga, famed knife fighter Jesús Cui, Eduardo Baculi, Nani Rabaya, Lucresio Albano, and my dad Vicente "Inting" Atillo. Several others were present during the gathering and became original members of the Balintawak Self Defense Club to include me as the youngest member at 14 years of age.

Original members of the Balintawak Self Defense Club shortly after it was founded in 1952. First row, first from left, Rabaya, second from left, Delfin N. Lopez, third from left, Venancio "Anciong" Bacon, fourth from left, Atty. Teodoro Arnoco, fifth from left, Atty. Eulalio E. Causing (1st elected President), sixth from left, Atty. Cecilio de la Victoria, ninth from left, Nani Rabaya, second row first from right, my father Vicente "Inting" Atillo, second from right, Timoteo E. Maranga, fifth from right, Jesús Cui, Sr., third row second from right, Eduardo Baculi, Sr., and third from right Lucresio Albano.

Styles and Systems

Just as there were different methods of eskrima that were practiced by the separate members of the Doce Pares Club, the same held true for the Balintawak Self Defense Club. When the club was founded, there were two distinct methods of eskrima that were predominantly being practiced by members of the club. One was that of Anciong, a student of Tatay Ensong, and the other was the style of his nephew Teodoro "Doring" Saavedra. This was the style taught by my dad, Vicente "Inting" Atillo, and Delfin Lopez. They, along with Anciong, were founding members of the club. In fact, many eskrimadors referred to this style as the *Mambaling Style* of eskrima.

Street sign at the corner of Balintawak Street and V. Gullas Street, Cebu City, Cebu, Philippines (2014).

In those days, *eskrima* was just *eskrima*. There weren't different "styles" of eskrima like today. An eskrimadors style of eskrima was determined by how the individual moved and the tactics they employed when sparring or fighting. Even though Anciong and Doring were training partners and students of Tatay Ensong, each of them developed their own way of doing things. This defined their particular method or style of eskrima. Anciong, my dad, and Delfin Lopez were close friends and training partners. However, my dad and Delfin Lopez were primarily students of Doring until his death. Therefore, they predominantly adopted Doring's style of eskrima. They, in turn, modified Doring's technique to reflect their own individual traits. Anciong, who was a student of Tatay Ensong along with Doring, became their primary instructor after the war. Just as two boxers from the same gym may train together under the same boxing coach, their individual styles of boxing will generally be different. One might be aggressive and have an aggressive style of boxing. The other may be a defensive fighter and counter-puncher. They both are learning under the tutelage of the same boxing coach. Still, each boxer will naturally have his own style and way of doing things, even though they are members of the same club.

Beginning of a Legendary Rivalry

Anciong's decision to formally leave the Doce Pares Club and form his own club rocked the Cebu eskrima community. He was candid and vocal about his disinterest in remaining a member of the post-war Doce Pares Club and his dislike for individual members of the Cañete family. Specifically, Ciriaco "Cacoy" Cañete. Anciong held Eulogio "Yoling" Cañete and Filemon "Momoy" Cañete in very high regard. They, in turn,

Balintawak Street today, Cebu City, Cebu, Philippines (2014).

recognized Anciong's superior skill and respected him as a peer. However, Anciong felt that Cacoy was a young, arrogant blowhard, and he would often challenge Cacoy to fight. Cacoy would never accept and was often protected by Yoling, who was the peacekeeper of the Cañete family.

Anciong was a very well respected and admired eskrimador. Several students and instructors of the Doce Pares Club followed him and joined the new Balintawak Self Defense Club. In addition to my dad and Delfin Lopez, other well-known eskrimadors left the Doce Pares Club to join Anciong. They included Timoteo "Timor" Maranga, Pio Deparine, Teodoro Arnoco, Cecilio de la Victoria, Nani Rabaya, Jesus Cui, Sr., Eduardo Baculi, Sr., Lucresio Albano, Lorenzo Gonzales, Isidro Barbilas, and Andres Olaibar.

Anciong's departure and the creation of the Balintawak Self Defense Club drove a wedge between Cebu's foremost eskrimadors. It incited an intense and legendary rivalry between the opposing clubs that lasted for the next several decades.

CHAPTER 6
Turbulent Era

The formation of the new Balintawak Self Defense Club marked the beginning of an intense rivalry. One that was intensified by local and national level political corruption and resulted in years of violent confrontation, sinister political agendas, and even murder. However, it was during this turbulent time that the art of Balintawak eskrima developed into one of the deadliest martial arts in the world. This period later became known as the Golden Age of Eskrima.

Cebu Politics and Delfin Lopez

After Sergio Osmeña, Sr.[77] of the Nacionalista Party,[78] lost the 1946 presidential election to Manuel Roxas of the Liberal Party,[79] Cebu was left in the hands of the Cuenco family. They were a well-known and influential political family. Their diverse business interests and political aspirations had been supported by national-level politicians in Manila. Much of which was used as a foil against Sergio Osmeña Sr. who had political control of Cebu before the war.

The family patriarch, Mariano Jesús Cuenco,[80] was the most anti-Osmeña politician in Cebu City and served as a member of the Philippine

---

[77] Sergio Osmeña Sr., (September 9, 1878 – October 19, 1961) was the patriarch of the Osmeña family and a founding member of the Nacionalista Party. Amid his accomplishments as a career politician, he served as the 4th President of the Philippines (August 1, 1944 – May 28, 1946), Vice President of the Philippines (November 15, 1935 – August 1, 1944), Senator of the Philippines from the 10th Senatorial District (1922 – 15 November, 1935), and Governor of Cebu (1904 – October 16, 1907).

[78] The Nationalista Party was founded in 1907 and is the oldest ruling political party in the Philippines.

[79] The Liberal Party of the Philippines was founded by senators Manuel Roxas, Elpidio Quirino, and form 9th Senatorial District Senator Jose Avelino on January 19, 1946 as a breakaway Liberal wing from the Nacionalista Party.

[80] Mariano Jesús Diosomito Cuenco (January 16, 1888 – February 25, 1964) was an attorney, politician, writer, and is the patriarch of the Cuenco family. The Cuenco family is a renowned

Senate as a member of Manuel L. Quezon's Commonwealth cabinet. His son Manuel Cuenco was the provincial governor of Cebu from 1946 - 1951.

Delfin N. Lopez, c. 1950.

During that time, the Cuenco family tightened their control over the local government and formed a close alliance with Philippine President Manual Roxas. However, on December 30, 1951, Sergio "Serging" Osmeña, Jr.[81], son of the former president, ousted Manuel Cuenco and was elected provincial governor of Cebu. This sent shockwaves through Cebu politics and reestablished Osmeña's political control of Cebu. Osmeña quickly embarked on an aggressive political campaign to remove Cuenco-supporting mayors around the province, alleging they were corrupt. This led to even higher tensions between the Cuenco and Osmeña families, both politically and for majority control of businesses in Cebu. Including the monopoly at the Port of Cebu. One such business interest of the Cuenco family was the Visayan Cebu Terminal Company, which had central control of the dockworkers responsible for handling cargo at the Port of Cebu.

The Shooting of Delfin Lopez

The bond between the Cuenco family and President Elpidio Quirino collapsed after the 1951 Senate elections when Mariano Jesús Cuenco lost his seat in the Philippine Senate to Quintin Paredes. The bond was further strained on April 9, 1953, when Quirino replaced Dr. Jose Rodriguez and appointed

---

political family who had substantial political and economic control of Cebu City. He was the 4th President of the Senate of the Philippines (February 21, 1949 – December 30, 1951), Senator of the Philippines (May 25, 1946 – December 30, 1951 / December 30, 1953 – February 25, 1964), Secretary of Public Works and Communications of the Philippines (1936–1939), Governor of Cebu (1931–1934), and Philippine House of Representatives from Cebu's 5th District (1912–1928).

[81] Sergio "Serging" Veloso Osmeña Jr., (December 4, 1916 – March 26, 1984) was a Senator of the Philippines, and is the son of Sergio Osmeña, Fourth President of the Philippines. On December 30, 1951, he was elected provincial governor of Cebu and Mayor of Cebu City for three terms in 1955, 1959 and 1963.

Vicente del Rosario as the major of Cebu City. Rodriquez had been serving as the mayor since July 24, 1952. His dismissal had been driven by political corruption and efforts by Quirino to gain control of Cebu City.

Venancio "Anciong" Bacon, Cebu City, Cebu, Philippines, c. 1950s.

Rosario immediately began campaigning for President Quirino, who was running for re-election in the 1953 Presidential Election against his former Secretary of National Defense, Ramón Magsaysay. Magsaysay resigned from his post on February 28, 1953, and defected from the Liberal Party to run for the Nacionalista Party. The Cuenco family abruptly switched their loyalty and began supporting the election of Magsaysay.

On May 11, 1953, Rosario further demonstrated his control of Cebu City by summarily dismissing forty-three detectives of the Cebu City Police Secret Service Force, including Delfin Lopez. He had become a member of the clandestine unit after the war. Rosario justified the terminations by claiming he had lost trust and confidence in the detectives. Their dismissal was arguably an attempt to manipulate the ranks within the Cebu City Police and remove detectives who posed a threat against the newly appointed Mayor and the re-election of President Quirino.

The termination of the detectives created a political backlash of resistance. Shortly after their dismissal, many of the detectives filed a lawsuit against the City of Cebu to reinstate them as detectives of the Cebu City Police Department, damages and costs, and back wages. The resistance was further fueled by the Cuenco family, who opposed the appointment of Rosario and the re-election of President Quirino. Delfin Lopez had become close to the Cuenco family and began threatening and intimidating the newly appointed mayor and his bodyguards. Each time a bodyguard was assigned to protect the Mayor, Lopez would threaten and physically assault them to show them he was in charge. Lopez fought with many of the bodyguards, and each time they were humiliated by the street-tough eskrimador.

Frustrated by Lopez and the lack of protection provided by the earlier bodyguards, the Cebu City Police assigned Vicente "Inting" Carin to protect Rosario. Carin was a long-time member of the Doce Pares Club and student of Filemon "Momoy" Cañete. As a policeman for the Cebu Police Department and fellow eskrimador, Carin was no stranger to Lopez. He was equally strict and had a reputation for being a dangerous fighter.

During a campaign event in the barangay of Tisa in Cebu City, Lopez and several former detectives of the Secret Service Force attempted to confront Rosario. He was at the event campaigning for the re-election of President Quirino. Upon his arrival, Lopez began antagonizing and provoking the Mayor and his bodyguards. Everyone, including the bodyguards, was intimidated by Lopez, and his presence at the campaign escalated the already heightened security concerns. Unintimidated, Carin confronted Lopez, at which time a heated argument ensued. Suddenly, Carin removed a Thompson submachine gun from beneath his jacket and began shooting at Lopez. Several bullets struck Lopez as he stumbled backward and fell through the railings of the Kinalumsan Bridge to the shallow river below. Fortunately, a tree beneath the bridge cushioned his fall as Lopez collapsed onto the muddy riverbank. In the chaos and anarchy that ensued after the shooting, several onlookers climbed down the riverbanks to render first aid to Lopez, who was taken to a nearby hospital where he narrowly survived the attack.

During the ensuing criminal investigation, witnesses claimed that Carin continued firing at Lopez even after he fell from the bridge. Several further stated that Lopez was not confronting the Mayor at the time of the shooting, and claimed the attack on Lopez was a politically charged attempt to assassinate him. Even more, witnesses believed the shooting was motivated by divided political loyalty and attempted to eliminate Lopez as the strong arm of the Cuenco family in Cebu. As a result, Carin was arrested and charged for the shooting and attempted murder of Delfin Lopez and incarcerated in Bilibid Prison in Muntinlupa, Philippines.

The incident greatly intensified the already hot-blooded tension between the Doce Pares Club and the Balintawak Self Defense Club. It marked the beginning of a turbulent era for Cebuano eskrima.

Venancio "Anciong" Bacon demonstrating at a local fiesta, c. 1960s. Anciong was rarely seen without a cigarette in his mouth even during his practice of eskrima.

Labor Union Strikes and Strike Busting

Shortly after leaving the Secret Service Force, Delfin Lopez founded his own security company, the Delfin N. Lopez Security Agency, and specialized in *strike busting*.[82] As labor disputes increased due to the unstable political and economic conditions in Cebu, so did the growth of labor unions and strikes by disgruntled union members. Lopez specialized in disrupting support for labor unions and the prevention of union strikes and the protection of company employees who were *strikebreakers*.[83] As a street-hardened and

---

[82] *Strike Busting*, or *Union Busting*, is a term commonly used to describe a wide range of legal and illegal activities carried out to disrupt or prevent the formation of trade unions and union strikes.

[83] A *strikebreaker* is a company employee or external person who works despite a union strike, or is employed to replace striking union members to ensure continued operations of a company, thereby making a strike ineffectual.

combat-experienced eskrimador, he would often use intimidation and physical violence to curb union support and deter the formation of strikes.

Lopez had very close ties with the Cuenco family and frequently worked for them as a bodyguard and provided other security-related services supporting businesses owned and operated by the Cuenco family. One such venture was the Visayan-Cebu Terminal Company. On the strength of the Cuenco family's political connections in Manila, the company attained renewal of its contract to manage cargo operations at the Port of Cebu in May 1956. This triggered a political backlash throughout Cebu and intensified the power struggle between the Cuenco and Osmeña families. Sergio Osmeña, Jr., son of the former President of the Philippines and the provincial governor who previously ousted Manuel Cuenco in 1951, began sponsoring efforts to sabotage operations of the Visayan-Cebu Terminal Company. Disgruntled dock workers, unhappy with the working conditions and divided by the opposing political factions, began to publicly outcry.

On April 17, 1954, Atty. Democrito T. Mendoza co-founded the Associated Labor Union (ALU) in Cebu City. He then launched the first ALU strike against the Visayan-Cebu Terminal Company in October 1956.[84]

Mendoza was a friend and student of Anciong and a former Balintawak Self Defense Club classmate of Lopez. The strike crippled operations of the Visayan-Cebu Terminal Company and paralyzed the port of Cebu. The Cuenco family hired Delfin Lopez to provide crucial strike-busting muscle and protect strikebreakers for the company so cargo operations could continue throughout the ALU strike. This created a problem for Mendoza, who had just been elected president of the ALU. Likewise, Lopez's support of the Cuenco family and efforts to deter the ALU strike created tension between him and another Balintawak Self Defense Club member, Atty. Jose Villasin[85], who was an attorney for the ALU, as well as Anciong himself, who was often hired to provide security at ALU strikes and events.

---

[84] "Labor Strike Paralyses Cebu Port Activities", *Manila Times*, October 1956, p. 16.

[85] Jose Villasin was a lawyer and officer of the Allied Labor Union (ALU) and member of the Balintawak Self Defense Club. He was the first President of Balintawak International Self-Defense Club, and is credited for organizing the curriculum of Venancio "Anciong" Bacon, and creating the Grouping method of Balintawak eskrima.

A few days later, Delfin Lopez was dubiously arrested by the Cebu City Police and jailed for a series of alleged murders dating back to 1947.[86] He was held in the Cebu Provincial Jail without bail, and his home was raided by the Cebu City Police under direct orders from Sergio Osmeña, Jr.[87] Osmeña then facilitated temporary control over dock operations in the Port of Cebu and recognized Mendoza and the ALU as the exclusive bargaining agent for all workers in the port. The arrest was made political "bail" by the opposing political factions. Lopez was eventually released, and all criminal charges were summarily dismissed.[88]

Delfin Lopez vs. Florencio Lasola

In 1956, a local eskrimador named, Otillo "Lolo" Larawan arranged a gathering of eskrimadors at his residence in the Lagtang *barangay* of Talisay City. Larawan wanted to eat lunch and participate in a friendly exchange of ideas and techniques. The invitation was extended to all eskrimadors in the area, including members of the rival Doce Pares Club. `I went to the gathering with Timoteo "Timor" Maranga, Edward Baculi, Delfin Lopez, Anciong Bacon, and several other members of the Balintawak Self Defense Club. I was surprised to see so many eskrimadors from the area. Eulogio "Yoling" and Filemon "Momoy" Cañete were there representing the Doce Pares Club along with Ciriaco "Cacoy" Cañete, Nicolas Javelosa, Saturnino Arcilla, Primo Albano, Tito Usurage, Fernando "Nanding" Candawan, Pedro Enrile, Mado Cañete, Luciano Cabañero, and Vicente "Inting" Carin. Carin had recently been released from Bilibid Prison in Muntinlupa for shooting Delfin Lopez.

Col. James Cushing, wartime guerrilla leader, dies on interisland vessel.

Death announcement of Lt. Col. James M. Cushing, August 26, 1963.

As soon as we arrived, I could feel that the tensions were high. Carin,

---

[86] "Ex-police Official Faces Murder Rap", *Manila Times*, 6 October 1956, p. 20

[87] "Cebu Murder Case Is Made Political "Bail" by Factions", *Manila Times*, 6 October 1956, p. 2

[88] Former Sleuth Will Sue Raiders", *Manila Times*, 3 November 1956, p. 5

who was representing the Doce Pares Club, had attempted to murder Lopez three years earlier, who was there to represent the Balintawak Self Defense Club. I also harbored hatred for Carin because of his attempted murder of Lopez, who was my godfather in the Catholic Church. The hostility between Carin and Lopez was worsened by the hatred and contempt Bacon felt toward Ciriaco "Cacoy" Cañete and the majority of the Cañete family. The friendly exchange Lolo had hoped for was rapidly turning into a clash between rival eskrimadors.

As expected, quarreling erupted between the two groups. Suddenly, Cacoy challenged both Anciong and Lopez. He stated, "If you are here to challenge the Doce Pares Club, I tell you both we never refuse a challenge hurled against us." Lopez immediately countered Cacoy and said, "We will fight any of you any time you want! Boxing, wrestling, judo, eskrima, whatever you want!" Suddenly, the crowd of onlookers became restless and started exchanging verbal insults and challenges. Cacoy was the most vocal of the Doce Pares Club members, yet he wouldn't commit to a fight. Finally, Lopez shouted, "You are talking too much, Cacoy! Stop talking and let's fight!" Cacoy evaded Doring's challenge and continued his bantering. Frustrated by Cacoy's bluff and nonstop antagonism, Anciong stepped in and challenged Cacoy himself. "Cacoy, I will fight you with only my stick, and I won't use my left hand!" Anciong was exceptionally skillful, using his left hand for *tapi-tapi*,[89] disarms, striking, and controlling his opponent. So exceptional, in fact, that use of his left hand had become very well known throughout Cebu.

Recognizing the confrontation was rapidly deteriorating, the elder Eulogio "Yoling" Cañete intervened to defuse the situation and instructed Cacoy and Anciong to calm down and not fight. Suddenly, Florencio "Inciong" Lasola, a very talented eskrimador who fought Cacoy earlier that year in a match that ended with them wrestling on the ground, challenged Lopez to fight. Lopez promptly accepted, and the crowd separated to make space for the two well-known eskrimadors.

As soon as the fight started, Lopez rushed forward and delivered three hard strikes to Lasola's head. Stunned by the blows, Lasola staggered

---

[89] Tapi-tapi is a method of hand checking wherein the free hand is used to rapidly check, or "tap" the limbs of an opponent to maintain control of and manipulate the opponent's hands and stick.

backward and crumbled to the ground. At that time, Lopez raised his stick once again to deliver a final lethal blow. Fearing that Lopez was going to kill Lasola, Anciong rushed in and disarmed Lopez's stick.

The crowd of onlookers became even more hostile and excited. To defuse the situation and protect Lasola from embarrassment, Anciong pushed the group back and helped Lasola to his feet. Anciong stated, "Inciong, your eskrima is lacking. I will practice with you slowly." Lasola was appreciative of Anciong's gesture but was embarrassed by the outcome of his fight with Lopez. Without hesitation, Anciong began practicing and drilling techniques with Lasola. Feeling that he needed to redeem himself in front of his peers, Lasola unexpectedly rushing in and disarming Anciong's stick. Shocked by Lasola's surprising and unexpected act, Anciong became angry. Disarming an eskrimador demonstrates superior skill and ability. It is often the deciding factor that determines the winner of a fight between two eskrimadors. Disarming an elder eskrimador in front of his peers, particularly one of Anciong's prominence, was tremendously insulting. Anciong was only trying to help Lasola and was not expecting Lasola's alarming action.

The group of eskrimadors and bystanders who observed the disarm were shocked. Irrespective of the rivalry between the Doce Pares Club and the Balintawak Self Defense Club, Anciong was very well respected by all eskrimadors. His skill was second to none. Lasola had disrespected Anciong in front of his students, peers, and rival members of the Cañete family. Lasola immediately recognized the consequences of what he had done and regretted his decision. Before Anciong could scold him, Lasola acknowledged his mistake and quickly withdrew into the gathering of eskrimadors. Anciong begrudgingly accepted the apology. The rest of the time at Lolo's residence was spent socializing and practicing with other eskrimadors without further incident.

Crispulo Atillo vs. Lauriano Sanchez

The most memorable experience of my life was my first challenge match in May 1964, when I was 26 years old. Lauriano "Lauren" Sanchez, a very tough and talented eskrimador from the Doce Pares Club, sent me a letter formally challenging me to fight. I eagerly accepted the challenge and agreed to fight him at the elementary school in the barangay of Tisa that following Sunday.

My fight against Lauren was my first official *deathmatch,* which meant there were very few rules. The protective equipment popular in arnis tournaments today had not yet been invented. The exact rules of each match were usually decided by the contestants moments before the match began. However, as challenge matches became more formal and the fear of legal repercussions increased, eskrimadors began drafting formal written agreements. Each participant would sign the agreement indicating they agreed upon the rules and released the other from any liability caused by physical injuries. My first cousin, Manual Gacho, draft the contract for the match, and it was presented to Lauren and myself shortly before the fight. The rules in Gacho's agreement specified that the match would consist of three rounds, and dropping a stick or disarm would immediately constitute an end to the round and loss of the respective round to the opponent.

A lot of people were there, including my dad and several family members and students from Mambaling. As the first round began, Lauren lunged forward, swinging his stick at my head, which I narrowly evaded and countered, hitting him in the leg. Lauren was wearing rubber knee-high fisherman's boots that absorbed the impact from my stick and protected his lower legs. Lauren continued his advance, at which time I struck him in the head. Momentarily stunned, Lauren dropped his stick, ending the first round. The second round began with Lauren advancing forward, wildly swinging his stick as I defended and moved backward. As I did, Lauren lost control of his stick a second time, forfeiting the round.

Beatriz Enriquez Atillo (b. July 29, 1946) married Crispulo "Ising" Atillo on December 10, 1962.

Frustrated and angry, Lauren turned his back to me and walked away. He was upset and requested that a new rule be adopted to allow the round to continue even if a fighter drops his stick. Lauren further asked that a circle be drawn on the ground around him and me to prevent one of us from retreating. If one of us stepped out of the ring, that fighter would lose the round. Lauren continued quarreling over the rules with my dad and refused to continue unless the rules were changed. Finally, the referee disqualified Lauren and declared me the winner.

Crispulo Atillo vs. Antonio Irogirog

In August 1964, Antonio Irogirog, Benjamin "Ben" Culanag, and several *goons*[90] unexpectedly showed up at my father-in-law's house in Punta Princesa. I lived there with my wife, Beatriz, who I married a few years earlier on December 10, 1962. She had given birth to our youngest son Rene on April 11, 1964. Irogirog was a member of the Doce Pares Club and was a student of Filemon Caburnay along with Lauriano "Lauren" Sanchez. I had easily defeated Sanchez in my first match six months earlier. Caburnay had been defeated by Delfin Lopez in a friendly challenge match a few years earlier in 1961. Culanag was considered one of Filemon "Momoy" Cañete's top students and his right-hand man. Irogirog and Culanag asked if Anciong or Delfin Lopez were at my residence. When I told them they were not, Irogirog abruptly challenged me to fight, which I readily accepted.

    A friend of mine, Manual Gocho, who was incidentally Irogirog's neighbor, found out that I had accepted Irogirog's challenge and warned me against fighting him. Gocho advised that he was friends with Irogirog and warned that Irogirog's *goons* would kill me if I participated in a fight against him. Gocho warned that Irogirog and his friends often carried guns and would most likely shoot me if I partook in the match. Especially if I won, which I was confident I would. I advised Gocho I was going to fight anyway, at which time he became angry and chopped my stick in half to prevent me from fighting. Determined to continue, I grabbed another stick and continued.

    Irogirog and I were presented with the agreement before the match and agreed it would consist of three rounds. We further agreed that if a fighter dropped his stick or was disarmed by his opponent, he would immediately lose the match.

    As soon as the fight began, I moved forward and landed several strikes to Irogirog's body. However, his footwork made him very difficult to hit. Irogirog had a "guerilla" or "hit-and-run" style and would move in and out very quickly. Suddenly, instead of running backward as he had done to

---

[90] A common term in the Philippines used to describe a person who is a bully or thug, especially one hired to intimidate others.

counter my previous attacks, Irogirog rushed forward and pushed me to the ground. As I fell back, Irogirog simultaneously struck my left ear with his stick. I immediately attempted to stand up and continue but was dizzy and disoriented from Irogirog's strike. That allowed Irogirog to disarm my stick, causing me to lose the match immediately.

A few days later, Delfin Lopez returned from Mindanao, where he had been working. Lopez found out that I had participated in the match against Irogirog and became very angry. He immediately confronted me about it and shouted, "Ising! Do not fight those people! They are our enemies and have a lot of guns! If you fight them and I am not there, they will kill you!" Lopez was not only upset with me for fighting Irogirog. He was also angry that Irogirog and the Doce Pares Club had come to my father-in-law's house to challenge me. Lopez believed they challenged me because he and Anciong were not there, and they were bullies. Lopez said, "Ising, we will train and go to Irogirog's house together and fight them when you are ready."

For the next several weeks, Lopez and I trained together to prepare for a rematch. Lopez stated when I was ready, we were going to do the same to them as they had done to me. He wanted to march up to their residences unexpectedly and challenge them to fight right then and there. Lopez was sure I could defeat them if everything were fair, and he was near to keep me safe. As tough and street-hardened as Lopez was, he possessed a strong sense of loyalty and was very protective of me as his godson. Unfortunately, we were not able to finish our training and carry out our plan to seek revenge against Irogirog and Benjamin "Ben" Culanag.

Death of Delfin Lopez

With control of the Port of Cebu, the ALU increased its power over the docks primarily through the systematic use of threats and violence. The ALU earned a well-deserved reputation for using strong-arm tactics to solicit union support and leading labor strikes.[91] On September 24, 1964, while attempting to calm an ALU labor strike in a rice warehouse owned and operated by the Uy Matiao & Company, Delfin Lopez was attacked by Emiliano Berania. Berania jumped from a tall pile of rice sacks inside the warehouse and landed on Lopez's back. Berania stabbed Lopez above his right clavicle with a long

---

[91] "PC Acts to Prevent Bloodshed at Waterfront Due to Shipping Strike", *The Daily News*, 22 March 1959, pp. 1, 2

knife that penetrated deep into his body. Lopez immediately collapsed to the floor and died before medical attention could be given.[92]

At the time of the attack, the ALU had staged a labor strike at the rice warehouse against the Uy Matiao & Company. They, in turn, hired Delfin Lopez and his company, the Delfin N. Lopez Security Agency, to calm the strike and protect *strikebreakers* and ensure production continued at the rice warehouse. It was alleged that Berania was hired by someone within the ALU to murder Lopez, and information concerning Lopez's location and time of arrival was deliberately given to Berania before the attack. It was further suggested that Berania purposely stabbed Lopez downward from behind because Lopez secretly wore a bulletproof vest that protected his chest and back. Not many people were aware of it. It was something Lopez started doing after he was shot by Vicente Carin of the Doce Pares Club a few years earlier. One of the few people who knew about the bulletproof vest was Anciong Bacon.

Delfin N. Lopez (d. September 24, 1964). This picture is framed above his remains in the Lopez family tomb.

Lopez's strike busting for the Quenco family and opposition against Mendoza and the ALU caused mounting tension between him and Anciong. Anciong worked as a laborer and was regularly hired by Mendoza to provide security at ALU strikes. Although Lopez and Anciong respected each other and had been long-time friends and training partners, tension had always existed between the two eskrimadors. Particularly after the death of Teodoro "Doring" Saavedra. Their friendship had become gradually more nonexistent and was worsened by Anciong's need for money and support of Mendoza and the ALU.

Delfin Lopez was a larger-than-life personality who was known by everyone. Word of his murder spread rapidly throughout the eskrima community, as did allegations of involvement. Growing political allegiances

---

[92] Ligayan Coban, "Immaculate Conception day murder in Mambaling", *The Freeman*, 15 December 1966, pp. 1, 14-15, 26

began to create contention between members of the Balintawak Self Defense Club, and instructors began to divide and form separate independent clubs. Conspiracy theories began to circulate, implicating several people throughout Cebu. Specifically, Anciong, Mendoza, and Sergio Osmeña, Jr.

The family tomb of Delfin N. Lopez at the Calamba Cemetery, Cebu City, Philippines.

I was aware of the long-standing issues between Anciong and Lopez. I witnessed first-hand the conflict between the two eskrimadors. Anciong was not present inside the warehouse at the time of the murder. Still, I believed the rumors that were circulating about his involvement. I felt he knowingly provided Berania with information about the bulletproof vest and where to stab Lopez. I wasn't the only one who believed that. Eskrimadors all over Cebu felt the same way as did my dad. He was closer than anyone to both Lopez and Anciong. Though I had known Anciong my entire life, and he was my grandfather's godson, the murder of Delfin Lopez signaled the end of my relationship with him.

The subsequent criminal trial of Berania was just as crooked and misleading as the murder itself. In a trial swayed by political intrigue, Berania was found not guilty and acquitted of the crime despite several witnesses who watched him stab Lopez. Most of the witnesses who testified were members of the ALU who condoned the attack. In contrast, others remained silent in fear of retaliation if they testify against Berania or implicated Mendoza and the ALU. Ironically, Berania was mysteriously killed in Mindanao, Philippines, not long after the trial.

Delfin Lopez was my godfather, friend, and one of my instructors. He was a decorated member of the Recognized Guerillas of USAFFE under the famed Lt. Col. James Cushing during WWII. He was a founding member of the Balintawak Self Defense Club. As an eskrimador, his exploits were legendary. He was the embodiment of a battle-hardened warrior. He was feared by those who opposed him and loved by those who cared for him. To the latter, he was a kind, protective, and generous man. To those who opposed him, he could be ruthless.

Delfin Lopez was interred in a small family tomb inside the Calamba Cemetery in Cebu City. The gated tomb remains to this day and is accessible to visitors wishing to visit and pay their respects to a Cebuano icon and legendary eskrimador.

Incarceration of Venancio Bacon

Not long after the murder of Delfin Lopez, Anciong was involved in a physical altercation, which ultimately landed him in prison for the next several years. Anciong's son, Meliton "Leony" Bacon, was a well-known robber and petty thief. He would steal just about anything to make money, and everyone knew of Leony's reputation as a thief. Leony had been stealing coconuts from the coconut trees on Pio Deiparine's property, who lived next door to Anciong. Deiparine was a friend and student of Anciong's and had been an original member of the Balintawak Self Defense Club since it was established in 1952. He had also been an original member of the Doce Pares Club since it was established in 1932.

Deiparine had grown tired of Leony's behavior and confronted Anciong from outside Anciong's window. Deiparine demanded that Anciong speak to his son and stop him from stealing things from his property. Anciong was inside his kitchen area at the time and, after concealing a knife in his waistband, came outside to confront Deiparine. Anciong and Deiparine began to quarrel over Leony's conduct and the theft of the coconuts. Anciong was furious because his son Leony had been arrested for an unrelated robbery a few days earlier and was incarcerated in jail. Deiparine was physically larger than Anciong and was skilled in both eskrima and boxing. The verbal argument between the two neighbor-eskrimadors escalated rapidly and turn into a physical confrontation. At that time, Anciong removed the knife he had concealed in his waistband and stabbed Deiparine in the stomach.[93] Deiparine immediately collapsed to the ground outside Anciong's home and bled to death.

When the police responded, Anciong claimed that Deiparine attacked him with the knife as he was walking home. He further claimed that he killed Deiparine in self-defense as he was attempting to defend himself. Unfortunately for Anciong, several neighbors in the area witnessed the confrontation. They provided the police with eyewitness accounts that did not

Venancio "Anciong" Bacon, c. 1964.

echo the sequence of events described by Anciong. The police concluded their investigation and, in the end, arrested Anciong for stabbing and killing Pio Deiparine.[94]

---

[93] Anonymous. *Personal Interview*. Name withheld at request of the witness due to fear of retaliation. Cebu City, Philippines, October 16, 2013.

[94] Atillo, Macario. *Personal Interview*. Cebu City, Philippines, October 14, 2013.

Several witnesses were subpoenaed by the prosecuting attorney to testify against Anciong during his trial, including my friend "Plesing" and Flora Alfar, the goddaughter of Pio Deiparine. Both testified under oath that Anciong stabbed Deiparine in the stomach and kicked him several times after he collapsed to the ground. Anciong's defense attorney argued that he disarmed Deiparine and stabbed him in self-defense because Deiparine was a much larger man skilled in eskrima and boxing. However, the prosecution echoed the witness testimony. It contended that Deiparine's death was not an act of self-defense and was murder. They further claimed that Anciong was a well-known expert in eskrima and should have used more considerable restraint against Deiparine.

Anciong was ultimately found guilty of murdering Pio Deparine and was incarcerated at the Criminal Detention Facility at Camp Crame,[95] headquarters of the Philippine Constabulary of the Armed Forces of the Philippines (AFP) in Quezon City. In an ironic twist of fate, his son Leony was killed a short time later while robbing a doctor's residence in the barangay of Labangon. The house was only a short distance away from where Anciong killed Deiparine.

Crispulo Atillo vs. Venancio Bacon

After Anciong was paroled from the Criminal Detention Facility at Camp Crame in 1975, he moved to Iligan City in Northern Mindanao. Anciong would often return to Cebu City as a condition of his parole and check up on his students. He would frequently join our training sessions in Mambaling behind the Lady of Lourdes Parrish Church[96] in Punta Princesa every week to practice.

During one of the gatherings shortly after his release from prison, Anciong wanted to demonstrate the innovations he had made while incarcerated. He also expressed that he was interested in reuniting everyone under his leadership as the chief instructor. I was there with my dad and several students to include Rudy Tabasa, Eduardo Tabasa, and Victor Tabar.

---

[95] Camp Rafael C. Crame is the national headquarters of the Philippine National Police (PNP) in Quezon City. It was formerly the headquarters of the Philippine Constabulary and Armed Forces of the Philippines (AFP).

[96] The Lady of Lourdes Parish Church is now called the Archdiocesan Shrine of Our Lady of Lourdes Church and is located on Francisco Llamas St, Cebu City, Philippines.

My uncle Macario Atillo[97] was also there, as was Timoteo "Timor" Maranga. Several former students of Anciong's had branched off during his incarceration and started their own clubs and added to the original eskrima taught by Anciong and the founding eskrimadors of the Balintawak Self Defense Club. Many of them didn't take his recent innovations or desire to unify the clubs together as the chief instructor seriously. That frustrated and angered Anciong, who unexpectedly challenged everyone to spar with him. Everyone was surprised and hesitant to accept the challenge. Anciong then gestured at me and said, "Ising! Spar with me!" Anciong was now 63 years old and no longer in his prime, but he was still an unrivaled and dangerous eskrimador. Eskrima is unique in that an eskrimador can continue practicing well past his youth and even continue improving. I was only 37 years old at the time and was still considered very young for an eskrimador.

Lady of Lourdes Parrish Church, Cebu City, Cebu, Philippines (2014).

Anciong and I began the first round peacefully and exchanged techniques back and forth. Target areas during a friendly sparring match are usually limited to the shoulders and legs. Anciong told me, "Ising! Watch your shoulders!" and I teasingly replied, "Anciong, you watch yours as well!" This was not an actual fight, so I held back considerably, as did Anciong. Each time Anciong hit my shoulder, I would counter and hit his back. Anciong started becoming more and more frustrated, and our informal match began to intensify. "Ising! Be careful!" my dad shouted as Anciong and I started taking the match much more seriously. The first round ended in an even score.

---

[97] Atillo, Macario. *Personal Interview. Witness to the match between Venancio "Anciong" Bacon and Crispulo "Ising" Atillo*. Cebu City, Philippines, October 14, 2013.

In between rounds, my dad said, "Calm down, Ising! This is only a friendly match. Do not fight him!" I wasn't trying to fight Anciong, but I certainly wasn't going to let him bully me. I respected Anciong, but this was one of the first times I had seen since his incarceration for murdering Pio Deiparine shortly after Delfin Lopez was killed. I harbored a lot of resentment toward Anciong and believed he was directly involved in my godfather's murder.

As the second round began, Anciong was angry and our friendly match immediately escalated into an actual confrontation. We started exchanging serious blows and no longer limited our strikes to each other's shoulders and legs. Now we were trying to hurt each other. My dad and the others watching recognized that the match had become serious and started shouting, "Stop! Stop! Don't fight!" At that moment, I smacked the cigarette out of Anciong's mouth and disarmed his stick as he attempted to counter-attack. The disarm meant an immediate loss of the round by standard rules. It infuriated Anciong, who picked up his stick and angrily gestured he wanted to continue. My dad immediately stepped in to separate us and stop the match from continuing.

The area behind Lady of Lourdes Parrish Church where the fight between Venancio "Anciong" Bacon and Crispulo "Ising" Atillo occurred (2014).

As I stepped backward to distance myself from Anciong, he thrust the point of his stick into my upper lip, causing it to split open and bleed. He was embarrassed and angry because I knocked the cigarette out of his mouth and disarmed his stick in front of the others watching. I became even more upset and immediately wanted to fight back. Still, my dad physically held me and prevented me from attacking Anciong. My dad was angry with Anciong as well and said, "Anciong! Why did you do that to Ising? The match had stopped!" My dad then turned to me and said, "Son, you have to respect

Anciong. He is family and the godson of your grandfather." I abruptly replied by saying, "No! He thrust his stick into my lip! He is not my teacher, and I will not respect him!" I was not only angry at Anciong for thrusting his stick at me. I also harbored a grudge toward him ever since he was suspected of being involved in the murder of my godfather and close friend Delfin Lopez in 1964. His lack of respect only worsened my bitterness and resentment toward him, and I now wanted to fight him in earnest.

I said to my dad, "Let us fight! I will win!" My dad shouted, "No, Ising! He is family, and you will show him respect!" In anger, I said, "If I lose to Anciong, I will quit practicing eskrima forever!" Anciong was eager to continue as well and shouted, "Let's continue, Ising! This time no limits!" Anciong's challenge meant he wanted to fight an actual *juego todo*[98] and not just a friendly challenge match, as we had initially agreed. Reluctantly, my dad stepped aside to give Anciong and I space to continue and said, "It is up to you."

As we continued, Anciong advanced forward in his normal open position and initiated his attack by swinging at my head, which I defended and countered. Anciong then kicked me in the leg, at which time I rushed forward and attacked. Anciong attempted to counter, but I trapped his stick with my left hand and began striking him repeatedly in the head. My dad shouted, "Ising stop!" as he and the other eskrimadors rushed in to separate us, ending the fight. I was furious with Anciong, and as we were separated, I said to him in anger, "If you weren't the godson of my grandfather, I would kill you!"

Following the match, my students rushed in to celebrate my defeat over Anciong. Even though it was unexpected, defeating Anciong in a spur-of-the-moment challenge match was a significant event. He was a legend in Cebuano eskrima and a master eskrimador. I had no intention of fighting Anciong at the get-together. Still, he openly challenged me in front of my students, and I harbored deep feelings of resentment toward him for the murder of Delfin Lopez. I undeniably respected him as one of the greatest eskrimadors in the history of Cebuano eskrima. I had known Anciong my entire life and had trained with him during practice sessions with my dad and

---

[98] Juego todo is a Spanish term that literally means, "game all." In eskrima it commonly refers to a match with "no rules" that is not limited by target areas and anything goes.

Delfin Lopez since I was a child. Many of which were before the Balintawak Self Defense Club was even officially created in 1952.

Nonetheless, challenge matches were common in eskrima. After the first round, it was mutually agreed that we would continue the match as a *juego todo*. As unexpected as it was, the unplanned match allowed me to avenge Delfin Lopez's murder and test my skill against Anciong. One of the greatest eskrimadors to have ever lived.

The Split of Balintawak

My fight with Anciong was only one of several reasons the Balintawak Self Defense Club formally disbanded and split between the core eskrimadors once loyal to Anciong. The original club had grown apart during Anciong's incarceration, and numerous independent clubs were already teaching Balintawak eskrima throughout Cebu. Each had its own head instructor and group of loyal students. A few of them existed before Anciong's incarceration. However, his frequent visits and tutelage before his imprisonment united the branches under one banner of Balintawak eskrima. In fact, the Balintawak Self Defense Club was never exclusive to the location on Balintawak Street in Cebu City, and I rarely practiced there. From the beginning, there were different locations where members would regularly get together and practice. My dad and I mainly practiced at the original location in Mambaling and Anciong and Delfin Lopez because it was close to our home. Others would gather at different locations and at the location on Balintawak Street when in Cebu City.

Venancio "Anciong" Bacon, Cebu City, Cebu, Philippines, c. 1980.

Anciong's imprisonment worsened the split between the various groups. It unintentionally forced each of them to grow into independent clubs with their own head instructor and respective style of Balintawak eskrima. Further complicating things, President Ferdinand Marcos declared Martial

Law a few years earlier on September 23, 1972, to subdue the rising civil unrest throughout the country and promote national development. This ultimately led to political corruption and a deterioration of the economic conditions throughout the Philippines. It caused a wave of violence and

Vicente "Inting" Atillo and Crispulo "Ising" Atillo, Cebu City, Cebu, Philippines, c. 1980.

human rights violations and led to widespread social unrest throughout the country. It also broadened the growing rivalry between several members of Balintawak eskrima. Many were hired by opposing political families, businesses, and the ALU to provide labor and security services.

    A few weeks after my fight with Anciong organized another get-together to demonstrate what he had developed during his confinement in Camp Crame. As before, most of those in attendance did not take Anciong seriously, which led to him being even more frustrated and angry with his former students. Particularly Jose "Joe" Villasin and Teofilo "Pilo" Velez, whose home served as the gathering place and headquarters of the Villasin-Velez branch of Balintawak eskrima. Anciong was always supportive of Villasin and respected him as an eskrimador and lawyer. However, Anciong was troubled by Villasin organizing Balintawak eskrima into a curriculum and creating the *grouping method*. He was suspicious of Villasin and felt that he had been president of the Balintawak Self Defense Club for too long. He feared Villasin was trying to gain complete control. Anciong eventually grew supportive of Villasin and the different clubs from the original Balintawak Self Defense Club. However, he was destitute and needed to teach eskrima to make money and provide for his family, especially after his release from prison.

During the get-together, Anciong wanted everyone to spar to determine who would be the next president of the Balintawak Self Defense Club. Whoever was victorious would become the new President. Similar to what occurred at the gathering behind the Lords Parish Church weeks before, the competitive sparring to determine the next president did not happen. However, Anciong's challenge communicated his desire to establish himself as the chief instructor of Balintawak eskrima and unite the various clubs together. Unfortunately for Anciong, most of his former disciples had already established their own independent clubs and were unwilling to formally unite. Though they all respected and recognized Anciong as the founder of Balintawak eskrima.

The eskrimadors that were once united together under the banner of the Balintawak Self Defense Club officially disbanded and separated themselves into five distinct groups. This was also the first time individual *styles* of Balintawak eskrima were formally recognized. Anciong organized Balintawak Orihinal and began teaching the Kuentada style of Balintawak eskrima he created during his incarceration at Camp Crame. José "Joe" Villasin, who founded the Balintawak International Self Defense Club, continued teaching his Grouping Method on Legaspi Street and his home in Lahug. Teofilo "Pilo" Velez continued teaching in his backyard on Sikatuna Street and eventually founded Teovel's Balintawak Arnis Group[99] on April 17, 1982. Timoteo "Timor" Maranga founded the Santo Nino Self Defense Club 19, which he later changed to Tres Personas Arnis de Mano.[100] Finally, my dad and I founded the New Arnis Confederation of the Visayas and

---

[99] The Teovel's Balintawak Arnis Group founded by Teofilo "Pilo" Velez is now the World Original Teovel's Balintawak Arnis Group (WOTBAG) and is headed by his sons Monie Velez, Pacito Velez and Eddie Velez.

[100] Timoteo "Timor" Maranga's eskrima is now called Combate Eskrima Maranga and is headed by his son Rodrigo "Drigo" Maranga in Cebu City, Philippines.

Mindanao (NAC), which we changed to the Philippine Arnis Confederation (PAC) on April 24, 1975.

Even though my dad was a founding member of the Balintawak Self Defense Club, he chose not to use the *Balintawak* name when we founded the PAC. He witnessed the politics and rivalries between his friends and former training partners that had developed over the years. Instead, he decided to use the term *arnis*, which had become popular following the founding of the National Arnis Association of the Philippines (NARAPHIL). However, he remained close friends with Anciong, and his decision not to formally use the *Balintawak* name was short-lived. As such, we later changed the name of the PAC to the Balintawak World Arnis Association. My dad proudly represented Balintawak eskrima and continued teaching until his death in 1993.

Even after Balintawak separated, Anciong still took the time to visit each club and socialize and give pointers when he felt it was necessary. He had hoped that his former disciples would come together and adopt the innovations he made while incarcerated. However, each had grown independently over time. He was friends with those who branched off and eventually became more supportive.

National Arnis Association of the Philippines

In January of 1975, General Fabian C. Ver, the Chief of Staff of the Philippines' Armed Forces, and Romeo C. Mascardo founded the National Arnis Association of the Philippines (NARAPHIL). The purpose was to unite the various clubs throughout the Philippines and revive the practice of *arnis* as a sport and means of self-defense. General Ver was appointed President, and Mascardo was selected to be the Executive Director of the association. The following year, NARAPHIL was accepted as an Associate Member of the Philippine Olympic Committee (POC)[101] and served as the first officially recognized national arnis organization in the Philippines.

Although NARAPHIL had good intentions, the association reflected the dubious politics of the Marcos Dictatorship. It was plagued by government bureaucracy and intrigue. Primarily that of General Ver. Ver

---

[101] NARAPHIL remained as the sole governing body for arnis until the government of the Philippines was completely revamped after the fall of the Marcos Dictatorship and People Power Revolution of 1986.

was feared throughout the Philippines and was notorious for taking no prisoners and resorting to torture when needed. In addition to NARAPHIL and his responsibilities as the Chief of the Armed Forces of the Philippines, Ver served as the Commanding Officer of the Presidential Security Command (PSC) responsible for providing security for the President of the Philippines, Ferdinand Marcos. Ver was also head of the National Intelligence and Security Authority (NISA), which Marcos established on September 16, 1972, to carry out covert and clandestine intelligence operations. Marcos used General Ver and NISA to track down and assassinate anti-Marcos antagonists.

General Fabian Ver (left), Presidential Guard Battalion Commander and Director of the Presidential Security Unit with former Philippine President Ferdinand Marcos (right).

Despite the politics and secrecy that plagued the Marcos Dictatorship, NARAPHIL was an enormous step forward for the art of eskrima, which was becoming more popularly known as *arnis*. For the first time, a sizeable governing body had been established to preserve and propagate the martial arts of the Philippines. The association rejuvenated interest in the art and encouraged a collaborative partnership between the numerous eskrima clubs and masters throughout the Philippines.

Shortly after NARAPHIL was established, I met with Romeo C. Mascardo, the Executive Director of NARAPHIL in Manila, Philippines, to discuss affiliating the New Arnis Confederation of the Visayas and Mindanao (NAC) with NARAPHIL. Mascardo approved my request without question and signed a membership certificate, making the NAC the first eskrima organization of Cebu to affiliate with NARAPHIL on April 24, 1975.

Cebu Eskrima Association

Not long after NARAPHIL was created, Dionisio "Diony" Cañete founded the Cebu Eskrima Association (CEA) in Cebu City in 1975. The association's purpose was to unite all the independent eskrima clubs on Cebu into a single association that could affiliate with the newly established NARAPHIL. Diony represented the Doce Pares Club and served as President, and José "Joe" V.

Villasin represented Balintawak eskrima as the Vice President of Administration. Ciriaco "Cacoy" Cañete of the Doce Pares Club served as Vice President of Operations. Other members of the CEA who represented Balintawak eskrima were Teofilo "Pilo" Velez, Atty. Sam Buot, and Johnny Chiuten.

Diony would often host meetings at Cacoy's house with senior members of both Balintawak eskrima and the Doce Pares Club. My dad and I would occasionally attend as did Venancio "Anciong" Bacon and Timoteo "Timor" Maranga. Even though Diony was sincere in his efforts to unite everyone under one roof, the bad blood and decades-old rivalry between the clubs and old-time eskrimadors was apparent. The meetings always seemed tense and on the threshold of erupting into violence. In addition to discussing the unification of all clubs under the CEA, we would often discuss the development of tournament rules and upcoming events. These discussions always lead to heated debates and disagreements and often resulted in eskrimadors hurling challenges at one another.

Officers of the Cebu Eskrima Association. Representing Balintawak eskrima were Atty. Jose "Joe" Villasin, Teofilo "Pilo" Velez, Atty. Sam Buot, and Johnny Chiuten.

The Doce Pares Club and Cañete family had a long-standing relationship with NARAPHIL. They were involved with it since its inception. In addition, the Cañete family had a venerable relationship with President Marcos and NARAPHIL president, General Fabian Ver. In fact, the

President's son, Ferdinand "Bongbong" Marcos Jr.,[102] was a long-time student of Cacoy Cañete.

In the beginning, Diony requested that each of us pay a membership fee of 200 pesos and agree to become members of the CEA. Most of the eskrimadors who attended the regular meetings agreed and paid Diony the money. I elected not to become a member because the New Arnis Confederation of the Visayas and Mindanao (NAC) was already affiliated with NARAPHIL. I was supportive of Diony and the CEA, but paying dues to affiliate with the association when our club was already a member of NARAPHIL seemed pointless. I instead chose to support the efforts of the CEA independently and retain my membership with NARAPHIL separately.

First National Open Arnis Championships

On March 24, 1979, the Cebu Eskrima Association (CEM), in association with NARAPHIL, sponsored the First National Open Arnis Championships in Cebu City at the Cebu Coliseum. The event was the first formally sanctioned tournament to implement the competition rules and protective equipment created by Diony to make full-contact competition safe.

Timoteo "Timor" Maranga, c. 1960.

I was there with Timoteo "Timor" Maranga to represent Balintawak eskrima as did two of my students, Alberto Ricablanca and William Torefill, who was also competing in the tournament. Maranga also brought students and was anxious to compete in the Master's Division and fight Ciriaco "Cacoy" Cañete.

At the time, I considered Cacoy a friend. Still, I was very aware of the politics associated with the Doce Pares Club. I advised Maranga that he would most likely be eliminated before facing Cacoy because of biased judging and

---

[102] Ferdinand "Bongbong" Marcos, Jr. is a Filipino politician and former senator in the 16th Congress of the Philippines. He is the second child and only son of former President Ferdinand E. Marcos.

prearranged tournament results. I reminded Maranga that the Doce Pares Club chose the judges and referees for the tournament. He had no experience sparring with the new protective equipment designed by Diony. Maranga was optimistic, however, and eager to face Cacoy.

Maranga's first fight was against Fernando "Nanding" Candawan of the Doce Pares Club. Candawan was an excellent eskrimador and was considered by many to be Cacoy's right-hand man at the time. Maranga convincingly defeated Candawan by disarms in all three rounds and advanced to the competition's second round. In his second fight, Maranga faced Benjamin "Ben" Culanag, a student of Filemon "Momoy" Cañete. It was Culanag and his friend Antonio Irogirog who challenged me to fight in 1964 when he and Irogirog unexpectedly showed up at my father-in-law's house in Punta Princesa along with several goons looking for Anciong and Delfin Lopez. I was anxious to see Maranga fight Culanag, and the winner would most likely face Cacoy in the finals.

As soon as the match began, Maranga disarmed Culanag's stick and quickly won the first round. During the second round, the two eskrimadors applied pressure to one another in an all-out back-and-forth exchange. Suddenly, Culanag wrapped his arm around Maranga's stick and attempted a disarm. Recognizing what Culanag was trying to do, I yelled out, "Maranga! Pull your stick!" Maranga immediately jerked his stick outward to counter the disarm, grabbed the other end behind Culanag's shoulder, and threw him to the floor. Culanag quickly recovered and began counterattacking. As Maranga blocked Culanag's strike, his stick made contact with Culanag's extended hand, which was a prohibited strike under the tournament rules. The referee immediately stopped the match and raised Culanag's hand in victory, claiming that Maranga committed a foul by striking his hand and was therefore disqualified. Culanag's win secured his position to fight for the championship and eliminated Maranga from the tournament.

After his disqualification, I reminded Maranga of what I said to him earlier regarding the biased judging and prearranged tournament results that would favor Cacoy. Maranga was angry. He was very aware of the Doce Pares Club's history of politics and cheating, but this was the first time he was personally impacted.

Cacoy's first match was against an eskrimador named "Hanco." Strangely enough, however, Hanco didn't report to the ring after the tournament officials announced his name, and he was consequently

Crispulo "Ising" Atillo, Cebu City, Cebu, Philippines, c. 1982.

disqualified. As a result, Cacoy was declared the winner by default and advanced to the next round of the competition to face Carlos Navarro of Black Eagle Eskrima in a semi-final match. After three consecutive victories over well-respected masters, Navarro was eager to fight Cacoy in the Master's division. Unlike other competitors, he declined to wear the protective padding on his hands and arms. After tournament officials announced Navarro would be facing Cacoy in the semi-final match, he was approached by two security officers who worked for Cacoy. Navarro was told that his and his family members' lives would be in danger if he participated in the match against Cacoy. Despite the threats to his safety, Navarro had rightly earned his position in the tournament standings and elected to continue and face Cacoy.

As soon as the match began, Navarro charged forward, pushing Cacoy backward with his aggressive style of eskrima. Cacoy attempted to defend himself but fell backward into the crowd as he retreated from Navarro. Navarro stopped momentarily to allow Cacoy time to recover and return to the ring. At that time, the referee unexpectedly stopped the match and declared Cacoy the winner. The crowd of spectators went crazy and began to chant, "White hair! White hair!" in support of Navarro, who had a complete head of grey hair. The spectators disagreed with the referee's decision. They felt that Navarro was the more aggressive fighter and was winning the match. Navarro was clearly upset by the decision and left the ring in disgust.

A short time later, tournament officials announced over the intercom that the final championship match would be between Cacoy and Benjamin "Ben" Culanag. Culanag immediately approached the tournament officials sitting ringside. He claimed he could not fight Cacoy because he was a student of Cacoy's uncle, Filemon "Momoy" Cañete, and represented the Doce Pares Club. After a brief discussion, the tournament officials agreed. The referee announced Cacoy was the winner of the tournament and declared him the National Champion. The crowd of spectators again erupted in disapproval. Although many supported Cacoy and the Doce Pares Club, most disagreed with declaring him the National Champion after only fighting in a single match. Particularly since that match ended in controversy. The controversial decision fueled everyone's suspicion that the tournament was fixed. It had been predetermined that Cacoy would be made the National Champion.

Before the First National Open Arnis Championships in Cebu City, Cacoy had boastfully claimed he had competed in over 100 deathmatches. However, none of us had ever seen him actually fight. Those of us representing Balintawak eskrima felt that Cacoy was not a real champion. We felt that declaring Cacoy the National Champion only reinforced his cheating reputation, worsening the long-standing rivalry between Balintawak eskrima and the Doce Pares Club.

First National Invitational Arnis Tournament

Five months after the First Open Arnis Championships in Cebu City, NARAPHIL sponsored the First National Invitational Arnis Tournament at the Philippine National College Gymnasium in Manila on August 19, 1979. Timoteo "Timor" Maranga had again been invited to compete in the Master's Division. I was there to support him and represent Balintawak eskrima.

In addition to Maranga, several other masters were invited to participate to include Ciriaco "Cacoy" Cañete, Arnulfo "Opong" Mongcal, José "Joe" Mena, Benjamin Luna Lema, Florentino "Tino" Pecate, Hortencio "Horton" Navales, and Antonio "Tatang" Ilustrísimo. Many of the masters

Crispulo "Ising" Atillo and student, Mambaling, Cebu City, Cebu, Philippines, c. 1982.

invited refused to compete under the tournament rules and felt the rules restricted their individual styles by imposing safety rules that made the competition unrealistic. Additionally, following the results of the First National Open Arnis Championships in Cebu City, Cacoy had earned a bad reputation throughout the eskrima community for cheating. Many of the competitors and masters in attendance believed that the Doce Pares Club had unfairly selected the tournament's judges and predetermined that Cacoy would win. Mostly since Cacoy was made the National Champion after competing in only one match that the referee had prematurely stopped. Despite the controversial results at the First National Open Arnis Championships, Maranga desperately sought to redeem himself and was anxious to fight Cacoy.

In Maranga's first and only fight, he faced Benjamin Luna Lema of Lightning Scientific Arnis. Luna Lema was a tough eskrimador originally from Mambusao, Capiz, on the island of Panay. He was also a veteran of the guerilla forces during WWII, with a karate background, *Combat Judo*, and bodybuilding. The fight turned out to be a very competitive and challenging match for both. During a fierce exchange of blows between the veteran eskrimadors, Maranga hit Luna Lema's head with his stick and, in the process, injured his right hand. The referee halted the fight to assess the extent of the injury. After a short pause, the referee declared Maranga medically unfit to continue and awarded the win to Luna Lema.

Cacoy fought Jose Mena, Arnulfo "Opong" Mongcal, and Florentino Picate to win the tournament and was awarded the Championship. All three fights lasted the entire three rounds and were decided by decision.

Death of Venancio Bacon

Toward the end of 1979, Anciong's health began to decline due to years of poor health and smoking. He continued innovating and teaching his *Kuentada* style of Balintawak eskrima and visit training sessions at various locations while offering advice from the sidelines. He continued teaching a loyal following of students to include José "Joe" Go, Arturo Sanchez, and others.

Another eskrimador who continued training with Anciong until his death was Bob Silver "Bobby" Tabimina of Iligan. Bobby began training with José "Joe" Villasin in 1967 and was introduced to Teofilo "Pilo" Velez a short time later. Bobby continued training with Velez until 1971 when he was able to meet Anciong, who at the time was still incarcerated at Camp Crame in Manila. Bobby began visiting Anciong regularly while imprisoned and remained a faithful student of Anciong's even after being paroled from prison.

The last time I saw Anciong was at a Cebu Eskrima Association meeting at Cacoy's residence a few months before Anciong passed away. His health had been deteriorating, and he looked frail. Nevertheless, even with failing health, Anciong was still a master eskrimador. He remained devoted to his students and eagerly participated in practice sessions and meetings to discuss eskrima until the end of his life.

During his final months, Anciong's family cared for him, as did many of his students. Unfortunately, years of impoverishment and smoking took its toll, and he passed away on November 1, 1980, at 69. Anciong's family buried him in a small public niche at the San Nicolas Catholic Cemetery on V. Rama Ave. in Cebu City. Sometime later, Octavius "Jimboy" Hife, a student of Anciong and Teddy Buot, had his remains exhumed and buried him with his wife Catalina "Lina" Decatoria Bacon in a proper stone ossuary where others can visit and pay their respects[103].

Anciong was undeniably one of the greatest eskrimadors who ever lived and an icon of Cebuano eskrima. As a student of Lorenzo "Tatay Ensong" Saavedra and childhood friend and training partner of Teodoro "Doring" Saavedra, Anciong stood out as one of the greatest fighters and innovators of our time. If not for the efforts of Anciong, the Balintawak Eskrima Club and Balintawak eskrima would not exist.

---

[103] Buot, Sam L. *Balintawak Eskrima*. Spring House, PA: Tambuli Media, 2014.

CHAPTER 7
The Last Challenge

For years following the First Open Arnis Championships in Cebu City and the First National Invitational Arnis Championships in Manila, Philippines, I repeatedly challenged Ciriaco "Cacoy" Cañete to fight. His controversial victories at the previous tournaments only fueled his public bantering and over-exaggerated claims of being undefeated and winning over a hundred *deathmatches*. He continued to publicly challenge other eskrimadors; however, he would actually never fight. The few eskrimadors who accepted his challenges were subjected to political pressure and threats and would ultimately be forced to withdraw. Cacoy would then boast publicly that his opponents were afraid to fight him and pulled out because they knew they would have been severely beaten or killed. Cacoy was without question a talented eskrimador, but his over-exaggerated claims and constant mocking disrespected other eskrimadors who were growing tired of his bantering.

I challenged Cacoy at every opportunity and offered to fight him anytime and anywhere. He would always respond by saying, "Ising, I am not afraid of you! I will fight you anytime!" Yet, when I would reply, "Let's fight right here and right now!" he would never actually follow through or formally accept my challenge. I eventually began to feel that Cacoy was not taking my challenges seriously and was intentionally avoiding me. I thought that I needed to publicly challenge him on behalf of all eskrimadors subjected to his constant belittling. In particular, those of us who represented Balintawak eskrima.

To force Cacoy to accept my challenge, I wrote a letter to the Sun Star Daily[104] newspaper in Kamagayan, Cebu City, publicly challenging him to a match. News spread rapidly throughout Cebu. It was immediately published in several other newspapers, including The Freeman, which headlined on July 1, 1983, that I challenged Cacoy to a fight without any

---

[104] The Sun Star Daily is the flagship newspaper of the Sun Star network of newspapers and is the leading newspaper of both Metro Cebu and the province of Cebu.

protective equipment "anytime, anywhere." I knew if the challenge were made public, Cacoy would be forced to accept otherwise face public embarrassment. He had boastfully claimed he would fight anyone willing to challenge him and was in the spotlight. My plan proved successful, and on July 3, 1983, the Freeman Daily published that Cacoy had finally accepted my challenge.

Even though President Ferdinand Marcos had lifted martial law in the Philippines a few years earlier on January 17, 1981, participating in a public duel was considered illegal. Additionally, following the creation of NARAPHIL by the Philippine government, the development of standardized tournament rules, and the use of the safety equipment created by Cacoy's nephew, Dionisio "Diony" Cañete, public *juego todo* matches between eskrimadors had mostly become a thing of the past. Consequently, it needed to be officially sanctioned for the match to occur, and we needed a permit from the Philippine Constabulary.[105] I was an instructor for the Philippine Constabulary, so I asked Major Antonio Medija, Commander of the Philippine Constabulary, to seek approval from Brigadier General Alfredo Olano, the Commander of the Region VII Headquarters. General Olano approved the match under the condition that it be considered an "eskrima exhibition" and not an actual *juego todo* match. A short time later, I received the permit we needed. The match was scheduled for 3:00 p.m. on Saturday, September 17, 1983, at the Philippine Constabulary Recreation Hall on Jones Avenue in Cebu City.

DYLA Radio Interview

Shortly after Cacoy accepted my challenge, he and I participated in a radio interview at the DYLA Radio Station in Cebu City to promote the fight. At the time of the interview, the match's specific rules had not yet been agreed upon. I proposed three rounds of 3-minutes each, and no protective safety equipment should be worn by either fighter. I also suggested that no other sport or martial arts such as judo or wrestling be allowed. The match is limited to strictly the

---

[105] The Philippine Constabulary was the national police force of the Armed Forces of the Philippines. Established in 1901 by the American colonial government to replace the Spanish colonial Guardia Civil, it was It was merged with the Integrated National Police to form the Philippine National Police on January 29, 1991.

Filipino martial art of eskrima. I was confident I could defeat Cacoy, who at the time was 63 years of age, and I was 45.

On the other hand, even though Cacoy was nineteen years older than me by the time he finally accepted my challenge, he was a very feared eskrimador. He was still considered the best fighter of the Doce Pares Club. He had been declared the National Champion in the Masters Division at the age of 60 just a few years earlier on March 24, 1979, at the First National Open Arnis Championships in Cebu City.

During the interview, the radio broadcaster, Ramonito del Rosario, asked Cacoy how long the fight would last. With complete confidence, Cacoy boastfully responded, stating the match would last "less than ten seconds" and that he would probably knock me out "within seconds." Cacoy's overconfidence infuriated me, and I was eager more than ever to fight him. I had been waiting a long time for this fight. I had grown tired of his endless bantering and erroneous claims of being victorious in over 100 deathmatches. Cacoy and I were friends before the First National Invitational Arnis Championships, and I always respected him as an eskrimador. However, soon after the championships, he began to publicly challenge me in front of other eskrimadors and challenge anyone affiliated with Balintawak eskrima. Despite the efforts of Cacoy's brother Diony to the contentions of the past and bring all eskrimadors together, Cacoy's challenges reignited the decades-old rivalry between Balintawak eskrima and the Doce Pares Club.

The Contract

I was very aware of Cacoy's history of *daya*[106] and the Doce Pares Club's politics. Even though Cacoy verbally agreed to the rules, I knew I needed to have a written contract drafted that defined the rules in detail and hold him accountable. I also wanted the contract to include a list of prohibited acts to ensure he adhered to the art of eskrima. Cacoy's style of eskrima, *Eskrido*[107], relied heavily on quick and rapid *corto kurbada*[108] and *abaniko* strikes as well

---

[106] Daya or magdaya is a Tagalog term that refers to someone who cheats or defrauds another through lies and deceit.

[107] A style of a martial art founded by Ciriaco "Cacoy" Cañete in 1951 that combines Doce Pares eskrima, Aikido, Ju-Jitsu, and Judo.

[108] Corto Kurbada is a combination of a Spanish loan word and Tagalog combined to describe the curved close quarter strikes used in the Doce Pares style of eskrima.

as grappling techniques from other martial arts to control and throw opponents. His knowledge of many different martial arts styles was never in question, only his over-exaggerated claims of winning over 100 deathmatches and his fighting ability as an eskrimador.

To formalize the rules, I asked Vivencio "Baby" Paez, the nephew of Delfin Lopez, to draft a contract that would document the match's rules and conditions. By signing the contract, both Cacoy and I agreed that anyone who violates any of the rules would be disqualified and considered the match's loser.

On Friday, September 16, 1983, I reviewed the contract drafted by Paez then gave him the signed original to be delivered to Cacoy. Paez, in turn, gave it to a member of the Philippine Constabulary who agreed to provide it to Cacoy. After reviewing the proposal and list of prohibited acts, Cacoy accepted the conditions and signed both sides of the document agreeing to the rules. Cacoy was confident he would win the fight regardless of the rules, even though they forced him to adhere to the Filipino Martial Art of eskrima.

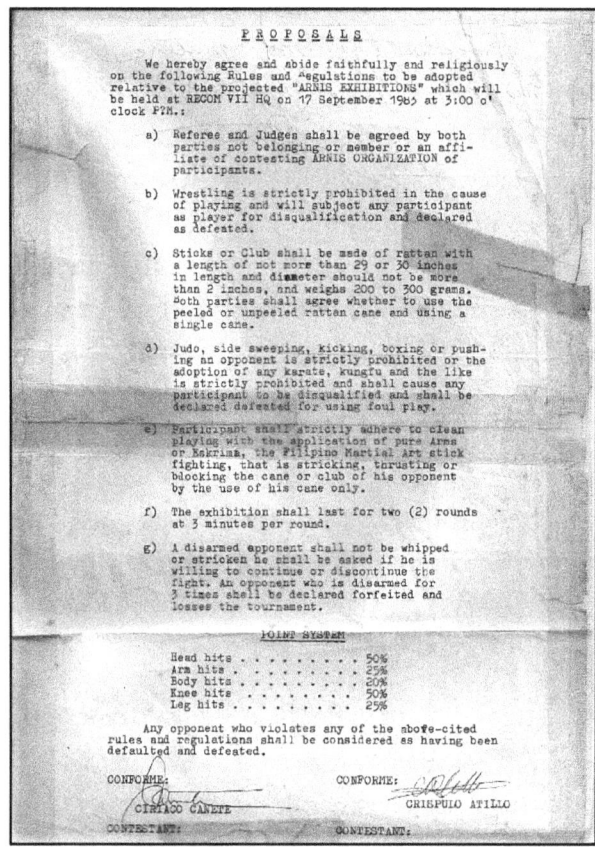

The original proposal signed by Ciriaco "Cacoy" Cañete and Crispulo "Ising" Atillo, September 16, 1983.

The Main Event

On the day of the match, the Recreation Hall of the Philippine Constabulary was packed with spectators and eskrimadors from all over the Philippines. The recreation hall was small and only held roughly one-hundred spectators; however, several hundred more anxiously gathered outside to listen to the match on loudspeakers. The match had been well advertised throughout the Philippines. People were eager to see two distinguished and well-known eskrimadors fight in a public event.

Cacoy was escorted to the ring by his older brother Eulogio "Yoling" Cañete and his younger nephew, Dionisio "Diony" Cañete. He entered the recreation hall from the adjacent locker rooms wearing a white Japanese karate *gi* with a black belt tied around his waist. Japanese-style karate uniforms and colored belts denoting rank were commonly used by the Doce Pares Club in training. Still, he didn't wear one at the previous tournaments, and we weren't sure why he was wearing one now. Those who wore the uniform were frequently ridiculed by traditional eskrimadors who were offended by Filipinos adopting Japanese culture and felt wearing the uniform was childish.

Camp Sergio Osmena Sr. Police Regional Office, Region 7 Headquarters of the Philippine Constabulary, Cebu City, Cebu, Philippines (2013).

Once Cacoy arrived in the ring, it was my turn to come out. I was escorted to the ring by my dad, Timoteo "Timor" Maranga, and José "Joe" Villasin. The match was a classic contest between the Doce Pares Club and Balintawak eskrima and probably the last of its kind.

As we came to the center of the ring to receive final instructions from the referee, Atty. Luciano Babiera, Cacoy, began to argue over the rules and refused to proceed until modifications were made. Cacoy demanded that the match starts in the center of the ring with our sticks crossed, and I wanted to start with both fighters on opposing corners of the ring. Eager to fight, I moved back to the right-hand corner of the ring and motioned to Cacoy to move back to his corner so we could begin. He again refused and remained standing in the center of the ring, arguing with the referee. Finally, frustrated and tired of Cacoy's quarreling over the rules and how the match should begin, Babiera instructed us to start from where we were standing.

Cacoy immediately attacked with his signature *abaniko* strikes, and I countered with a hard, vertical counterstrike to Cacoy's forehead. I felt the impact of my stick on Cacoy's head, at which time he lunged forward and grabbed me in a headlock. I attempted to break free, but both of Cacoy's arms were wrapped around my neck, and he refused to let go. I then grabbed Cacoy's right elbow to control his stick with my left hand and began striking the outside of his left leg. As my stick hit his leg, I realized why Cacoy was wearing the karate *gi*. He was wearing protective padding underneath his karate uniform that was protecting him from my strikes. I then began striking the outside of Cacoy's left ankle, which was exposed and unprotected by padding. Each time my stick impacted his ankle, I could feel Cacoy buckle and grimace from the pain as he desperately continued to hold me in the headlock. In an attempt to get Babiera to stop the match, Cacoy began shouting, "Referee! Referee!". Suddenly, Captain "Jojo" Yap of the Philippine Constabulary, who was standing ringside, and one of Cacoy's students, Sgt. Magdamit rushed in and disarmed my stick while I was still striking the outside of Cacoy's ankle. Still refused to let go, Cacoy held on tightly until the crowd of spectators swarmed in and physically separated us.

As Babiera attempted to regain control of the situation, accusations and counter-accusations began to fly as each of us claimed the other had violated the rules. I had inflicted a lot of damage to Cacoy's ankle. I believed he was going to be disqualified for wrestling and wearing protective padding underneath his uniform. I raised my hands in victory and began to celebrate with my dad, Maranga and Villasin, as well as nearly everyone watching from ringside. Cacoy continued arguing with the referee, claiming I was trying to tackle him. Obviously, everyone at ringside observed I was not because I was striking Cacoy's ankle as he held onto me.

Realizing he had violated the rules and was facing disqualification and public embarrassment, Cacoy pleaded with Babiera and begged him to start the fight over. We felt there was no reason and argued that Cacoy violated the rules and was therefore disqualified. He had also been caught cheating by wearing protective padding underneath his uniform.

Cacoy continued arguing and protesting the match results for several more minutes as the crowd became increasingly restless. The fight ended

Crispulo "Ising" Atillo vs Ciriaco "Cacoy" Cañete, September 16, 1983, Cebu City, Philippines.

quickly, and the spectators watching inside the recreation hall were at odds. Those supportive of Balintawak eskrima celebrated as others complained the match ended too soon. Feeling there was no reason to remain during the post-fight confusion, my dad, Maranga, Villasin, and I exited the ring to return to the locker room and celebrate. The outcome was clear to those of us representing Balintawak eskrima. Cacoy had lost, and the match was over.

Suddenly, feeling the pressure from Cacoy and the quarreling spectators, Babiera ruled that each of us had committed a foul. Through the commotion and noise of the crowd, Babiera instructed me to return to the ring so he could start the match over as Cacoy had desperately requested. Nevertheless, we had already left. Once we received word to return to the ring, we were stunned and argued that several minutes had passed, and the match was already over. As the back-and-forth quarreling continued, instead of disqualifying Cacoy for clearly violating the rules, Babiera raised Cacoy's hand and declared him the winner. The crowd of spectators went crazy, and just as we had done several minutes earlier, Cacoy and the representing members of the Doce Pares Club began to celebrate. Instead of protesting Babiera's outlandish decision, we recognized that we had again been cheated by Cacoy and subjected to the political influence of the Doce Pares Club.

Recognizing the outcome of the fight was going to be controversial, Timoteo Maranga quickly retrieved the original contract from the Philippine Constabulary and returned it to me for safekeeping. Maranga was also sure Cacoy had lost the match and would be disqualified when we left the ring. He recognized we were not going to be able to change the outcome but said, "Keep this Ising. This is proof Cacoy broke the rules and cheated."

Aftermath and Rematch

After I returned home, I was contacted by my friend, Loloy Uy. Uy stated that several radio stations reported that I had been severely beaten by Cacoy during the match. They further broadcasted that I was seriously injured and was transported to the hospital. Uy advised he went to the hospital to visit me, but when he arrived, I was obviously not there. The story was clearly a lie and being used as propaganda for Cacoy. I immediately went to DYLA Radio with Molo Rosada, one of Timoteo Maranga's top students, to set the record straight.

When we arrived, the radio announcer, Naving Velez, was surprised to see me and immediately acknowledged that I wasn't injured at all. I even lifted my shirt up to show Naving that I was not injured in any way. I advised Naving that I had not been hit by Cacoy at all during the match. When I asked where he got his information, he stated he received a telephone call from an unknown caller immediately after the match who advised him that I had been severely beaten by Cacoy during the fight. To settle the dispute, Naving telephoned Cacoy at home while broadcasting the conversation live on the air. Naving advised Cacoy that I was at the radio station to prove I had not been injured, and the earlier broadcast announcing I was in the hospital was not true. Cacoy acted surprised by the phone call and was hesitant to answer any questions over live radio.

Cacoy asked, "Ising, are you there? I hit you on the left and right temple." I immediately laughed and said, "Cacoy, you are a liar! You never him me! Not even once! I hit you four or five times, then you started wrestling!" I then challenged Cacoy to come down to the radio station so he could see for himself. Cacoy cautiously asked, "Why do you want me to go, Ising? What are you going to do if I come down to the station? Are you going to shoot me or kill me?" I said, "No, Cacoy, I am a sportsman. I only want to prove I wasn't injured, and you are a liar!"

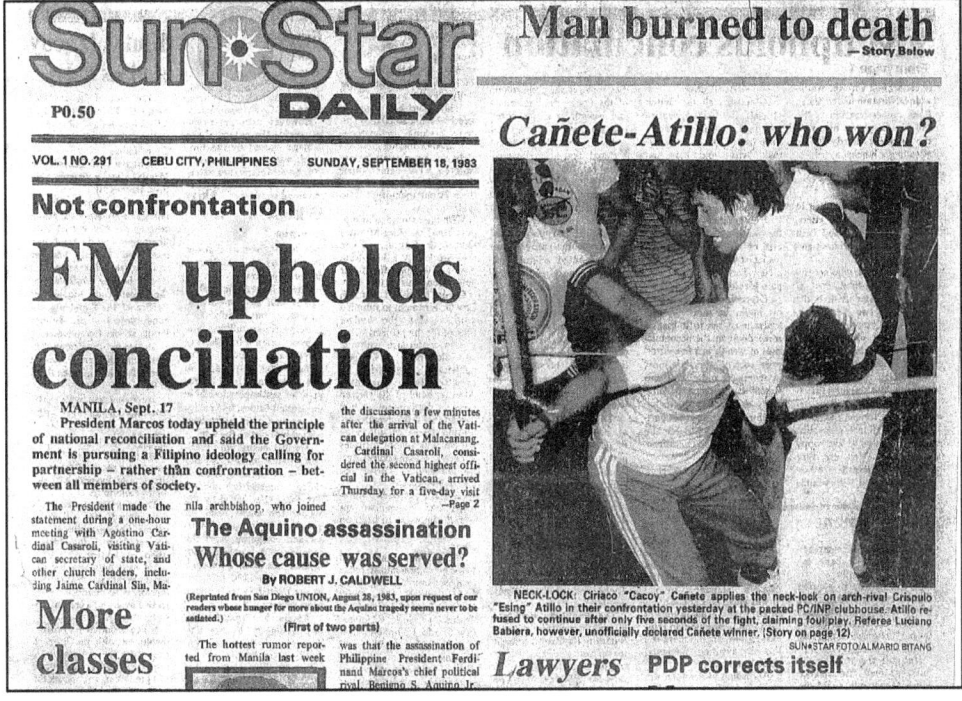

The Sun Star Daily newspaper, Sunday, September 18, 1983.

I advised Cacoy that if he came down to the radio station, I would have a neutral doctor examine us both for injuries and broadcast the live radio results. I told him, "Cacoy if the doctor finds any injuries on me, I will publicly admit that you beat me. But, if the doctor finds any injuries on you, you must also admit defeat." Naving interrupted and asked, "Cacoy, you said you won the fight, but the rules say that if anyone wrestles, they would be disqualified." Cacoy refused to answer Naving's question over live radio and responded by asking, "How much money did Ising give you to say that?"

After quarreling for several minutes over live radio, Cacoy refused to come down to the radio station and continued to deny any violation of the rules or admit that he had lost the fight. As the telephone call ended, Cacoy and I agreed to a rematch in December. I added, "Don't forget, Cacoy. When you go to sleep at night, you will always remember that you never hit me, but I hit you, and when you wake up every morning, you will always know that you are a liar, and I won the fight."

On Sunday, September 18, 1983, the Sun Star Daily published an article titled "Cañete - Atillo: Who won?" The paper even displayed a cover photo of Cacoy clearly violating the rules by executing a "neck lock" while I was striking his ankle. The article acknowledges that Cacoy broke the rules and states that I refused to continue because Cacoy committed a foul and because the referee, Luciano Babiera, "unofficially," declared Cacoy the winner. The decision was considered unofficial because everybody knew that Cacoy violated the rules. The fight had already ended by the time Babiera granted Cacoy's request to restart the match.

Even until Cacoy's unfortunate death on February 5, 2016, he always maintained that I grabbed his legs during the match, and he countered with a headlock to prevent me from tackling him. That is obviously not true and visible in the photo published on the Sun Star Daily cover the following day. I am not holding Cacoy's legs in any way and am striking his left ankle with my stick in the picture. Both of his arms are wrapped around my neck to stop me from hitting him. It's evident because his stick-hand is beneath my neck, and his stick is on the right side of my head as I'm shown striking his ankle.

Missing Video Footage

I left DYLA Radio and immediately went to DYKC-TV Channel 9 to retrieve a copy of the video footage they recorded of the fight. When I arrived, I was told by the station attendant that the only video footage of the match had already been taken by Cacoy's nephew, Dionisio "Diony" Cañete. I then went to Channel DYCB-TV Channel 3 and DYSS-TV Channel 7, who also recorded the match. Both stated that a member of the Cañete family retrieved the only recorded footage of the fight shortly after the match ended. To this day, nobody has been able to find the missing video footage.

The Rematch

Three days later, on September 20, 1983, I was summoned back to the Police Regional Headquarters of the Philippine Constabulary at Camp Sergio Osmeña in Cebu City to participate in a meeting. I wasn't sure what to expect.

Under the regime of then Philippine President Ferdinand Marcos, hundreds of people went missing, and thousands were imprisoned and killed. Many of these murders were ordered and carried out by General Fabian Ver, President of the National Arnis Association of the Philippines (NARAPHIL) and Commanding Officer of the Armed Forces of the Philippines under President Marcos. The Cañete family had a long-standing relationship with both the President and General Ver. They had expressed a keen interest in my match with Ciriaco "Cacoy" Cañete only days earlier.

Former Philippine Senator, Benigno "Ninoy" Aquino Jr.

Additionally, four weeks earlier, on August 21, 1983, former Philippine Senator Benigno "Ninoy" Aquino Jr. had been assassinated moments after landing at the Manila International Airport following a three-year exile in the United States. Aquino was a longtime political opponent of President Ferdinand Marcos. The public strongly suspected Marcos and Ver of the assassination.

Fearing that something may happen, I went with my dad, Vicente "Inting" Atillo. I was joined by Timoteo "Timor" Maranga when we reached the camp. Maranga was a Major with the Cebu Police Force and was also ordered to report to the Constabulary and participate in the meeting. None of us knew what to expect, and we were nervous.

Upon arrival, we were greeted by my friend, Col. Luis Kintanar, Judge Advocate of the Philippine Constabulary, and quickly shuttled into a small private room away from public view. Once inside, we were surprised to see Cacoy and Diony sitting in the office waiting for us. Kintanar explained that several anti-Marcos demonstrations and protest marches were scheduled to occur around Cebu City the next day and converge at a rally at Plaza Independencia. The rally had been planned by anti-Marcos activists to protest against the assassination of Senator "Ninoy" Aquino and demand an end to the Marcos dictatorship. Kintanar advised that the Region VII Commander of the Philippine Constabulary, Brigadier General Olano, received direct orders from the Office of the President instructing him to create a plan that would prevent the marches from occurring and divert the anti-Marcos protestors from the rally.

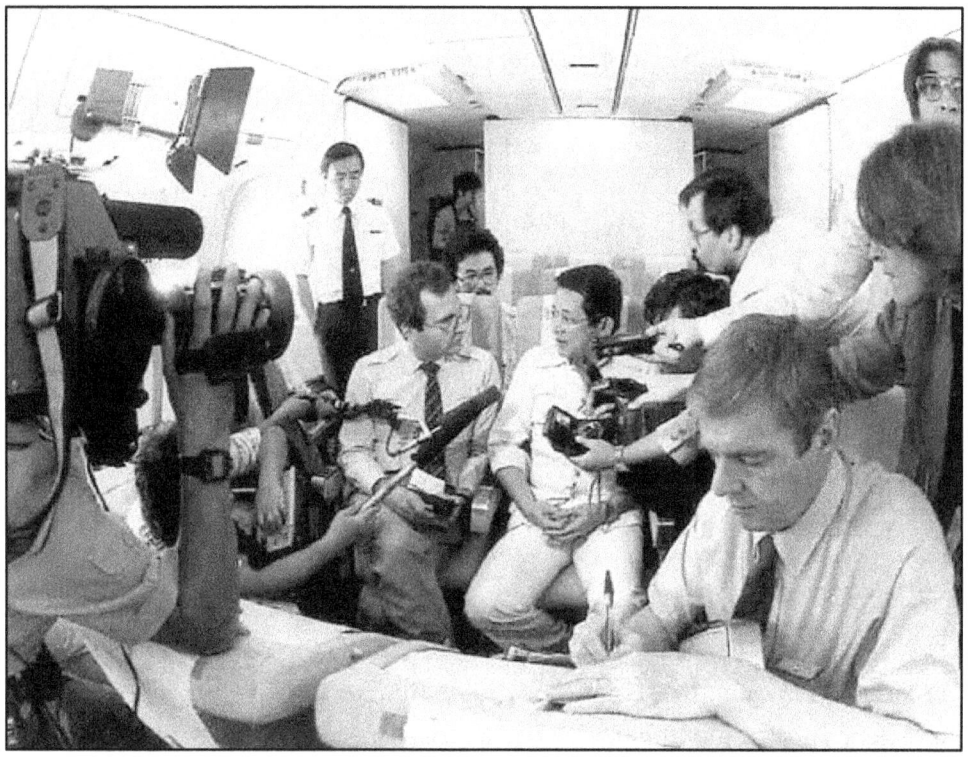

Former Philippine Senator, Benigno "Ninoy" Aquino Jr, being interviewed onboard China Airlines Flight 811 just moments before his assassination on August 21, 1983, at the Manila International Airport, Manila, Philippines.

Kintanar stated the fight between Cacoy and me the previous week had gained a great deal of media attention. Kintanar explained that the Philippine Constabulary had conceived a plan to use a rematch to divert protestors from the rally at Plaza Independencia and subdue the anti-Marcos demonstrations. Kintanar suggested that a rematch between Cacoy and I would draw demonstrators away from the rally, and preliminary matches before the main event would keep them away from Plaza Independencia long enough for the demonstrations to subside. Kintanar stated that even though the rematch would distract protesters from the rally, Cacoy and I would not actually fight. Instead, after several preliminary matches, Cacoy and I would be called to the center of the ring by the referee, where a doctor would publicly examine each of us. The doctor would then declare one of us unfit to fight, and the match would be postponed. Kintanar stated it had been decided during a discussion with Cacoy that the doctor would diagnose me with high blood pressure. He would then publicly announce that I was physically unfit to participate in the match.

I became angry and asked, "Why do I have to be the one found unfit to fight? Why can't we just fight?" Without providing an explanation, Kintanar snapped, "No, Ising! You will do it! The doctor will say you have high blood pressure, and the fight will be called off at the last minute."

I understood the Constabulary's plan to use the rematch to lure protestors away from the rally. Still, I did not agree with the decision to find me unfit to fight. I didn't have a single injury from our match earlier in the week. I suspected this to be another one of Cacoy's deceitful ploys. I argued back and shouted, "Cacoy was the one injured! He can be the one found unfit to fight! Not me! Have someone else do it!" I desperately wanted to fight Cacoy in a legitimate rematch. I did not want my reputation smeared by participating in a ruse. Especially one that suggested I was unfit to fight because of injuries allegedly received from Cacoy, which wasn't true. I said, "Let Maranga fight instead of me." Recognizing the potential danger and repercussions he would likely face if he participated in a match with Cacoy, particularly a ruse to stifle an anti-Marcos demonstration after the assassination of Senator "Ninoy" Aquino, Maranga declined and stated, "I won't fight Cacoy, Ising. I want to live!"

Cacoy sat quietly and didn't speak, which was not like him. He was usually very opinionated and vocal about matters; however, he refused to say anything this time. His silence only confirmed my suspicion he was somehow involved.

Kintanar asked to speak to me in a separate room so we could talk in private. Once inside, Kintanar said, "Ising, please just help us. You are an instructor for the Philippine Constabulary, and we have been ordered to carry out this plan." Kintanar explained that the Philippine Constabulary had already discussed the proposal with Cacoy, and it was Cacoy who suggested that I be found unfit to fight. Kintanar stated, "Ising, you don't have any visible injuries, so you have to be diagnosed with high blood pressure. Cacoy has a great deal of influence, and the decision has already been made." Kintanar warned, "Ising if you don't go along with the plan, your life could be in danger."

Kintanar and I then rejoined everyone else sitting in the other room. Kintanar advised everybody that the rematch would occur the following day at 3:00 p.m. at the Cebu Coliseum. He then reminded everyone that the ruse details were confidential and ordered us not to speak to anyone about the plan. Kintanar stressed that doing so would likely lead to severe consequences. Everyone agreed and shook their heads obediently, acknowledging they understood. Infuriated and feeling manipulated, I stood up and stormed out of the room to leave. As we neared the exit of the camp, a police officer confronted me. He quietly warned that I would be shot if I didn't show up at the Cebu Coliseum and strongly recommended that I go along with the plan.

Once I was safely home, I discussed the meeting with my dad and Maranga, who urged me to show up at the Cebu Coliseum at the time scheduled and go through with the plan. Otherwise, they both feared my life would be in danger, as did my wife, who was afraid I would be killed.

Crispulo "Ising" Atillo and his father, the late Vicente "Inting" Atillo, Mambaling, Cebu City, Philippines, c. 1983.

That night I couldn't sleep. I wanted to fight Cacoy in a rematch more than anything but was afraid my family would be harmed if I didn't go along with the ruse. I tried to imagine what would happen if I refused the doctor's ruling and instead demanded to fight in front of a stadium full of spectators. I also tried to imagine what would happen if I simply didn't show up. The more I tried to figure out a solution, the angrier I became at Cacoy. Declaring one of us unfit to fight wasn't necessary. An actual rematch would have sufficed the Constabulary's plan to lure protestors away from the political rally at Plaza Independencia and subdue the anti-Marcos demonstrations. It was Cacoy himself who came up with the idea to have me examined by a doctor and found physically unfit to fight. I had challenged Cacoy just days before on live radio. I even offered to have a neutral doctor examine both of us for injuries. He refused to participate because he knew he was the only one injured during our match. Before the meeting at Camp Sergio Osmeña, Cacoy used his political influence to deliberately devise a scheme that would enable him to save face and publicly discredit me. It was his way of regaining his credibility following his loss during our match the previous week and ensuring he and I would never fight again.

The next morning, I was again confronted by several policemen from the Philippine Constabulary near my home. They warned that if I did not show up for the event that afternoon or go along with the plan, someone would throw acid in my face, and I would be shot. The warnings were relentless, and I still wasn't sure what to do. I again discussed my options with my dad and wife. I realized I had no other choice but to go along with the ruse. I left my home in Mambaling and began making my way to the Cebu Coliseum. I also chose to go alone and instructed my dad to stay home where it was safe. I was still afraid I would be killed, or my dad would be physically harmed.

The Cebu Coliseum

When I arrived at the Cebu Coliseum, the arena was packed with hundreds of spectators excited to see the rematch. Numerous anti-Marcos protestors were still marching to the rally at Plaza Independencia. Still, the plan was successful, and hundreds of demonstrators had been diverted. As I entered the coliseum to prepare for the match, I was confronted by several policemen from the Philippine Constabulary. They once again warned me that acid would be thrown in my face if I did not go along with the plan, and I would be shot.

After several preliminary matches, the main event between Cacoy and I was scheduled to begin. Like the fight the weekend before, Cacoy was escorted to the ring by his brothers and entered the ring wearing his signature white Japanese karate-style gi and black belt. This time, however, he was limping and being helped to the ring[109]. It was obvious his ankle had been injured the week before. I was escorted to the ring by Col. Kintanar and several Cebu City police officers who had been assigned by the Philippine Constabulary to protect me during the event. Their job was also to ensure I participated in the ruse and didn't do anything that would compromise the event. I was afraid I was going to be shot regardless of my reluctant

---

[109] Anonymous. Personal Interview. Name withheld at request of the witness due to fear of retaliation. Cebu City, Philippines, April 20, 2014.

participation. I was also fearful that my dad would be harmed, so it was decided before the match that he and Maranga would not attend the event.

Cebu Coliseum, Cebu City, Philippines, 2014.

It was then announced by the ring-announcer that Cacoy and I would be examined by Dr. Eduardo Tojong. He had been hired by the Constabulary to perform the pre-fight medical examinations. A doctor's physical exam in the middle of the ring before a fight was highly unusual, and the crowd immediately suspected something scandalous was in the works. As had been previously decided by the military, Dr. Tojong examined Cacoy and announced he was physically fit to fight. The doctor then examined me. He concurred I didn't have any physical injuries but announced that I was suffering from high blood pressure. Dr. Tojong then asked me if I wanted to continue and fight anyway. I wasn't sure what to say. The option to fight Cacoy wasn't part of the plan, and I had been warned I would be in danger if I didn't go along with the ruse and do precisely what I was told. I desperately wanted to fight Cacoy again. But then also, I was afraid acid was doing to be thrown in my face, and I would be shot if I went against the Constabulary's plan. Fearing for my life and feeling I had no other choice but to protect my family, I reluctantly replied, "No, I cannot fight."

The crowd of spectators immediately erupted in protest and began throwing bottles and trash into the ring. They suspected a scandal had occurred and were furious. I was covered and protected by Col. Kintanar and the police officers assigned to protect me and was immediately escorted out of the ring and back into the locker room. Fearing I still wasn't safe, I quickly packed my belongings and left the Cebu Coliseum.

Aftermath

The next day I was contacted by Major Antonio Medija of the Philippine Constabulary. He instructed me to return to the Police Regional Headquarters at Camp Sergio Osmeña and report to his office. As soon as I arrived, Medija handed me an envelope containing 500 pesos. "What's this?" I asked. Medija stated, "This is from Diony. It's a percentage of the ticket sales from the match yesterday."

The money from Diony was appreciated, but it confirmed my suspicion that the Cañetes were closely involved in the ruse and were the principal organizers of the event. Not only did Cacoy not have to fight me in a rematch, but the Cañetes also profited from the sale of tickets for the event. All of this was a ruse led by the Philippine Constabulary, a branch of the Philippines' Armed Forces under President Marcos.

I took the warning from Col. Luis Kintanar very seriously and rarely spoke of the ruse after the event. The Philippines was a dangerous and violent place under the dictatorship of President Ferdinand Marcos, and I was often warned that if I told anyone, I would be killed.

Ongoing Challenges and Failed Rematches

Over the years that followed, Cacoy continued to boastfully claim he was the best eskrimador in the world, and I was afraid to fight him. That definitely was never the case. He was well aware of the reasons I was forced to withdraw from our rematch on September 21, 1983. He was personally involved in planning the ruse with the Philippine Constabulary and orchestrated the entire ruse. In fact, he and I frequently argued about it throughout the years. His arrogant claim that I was too afraid to fight was a blatant lie that he knowingly spread to falsely promote himself as the greatest eskrimador in the world.

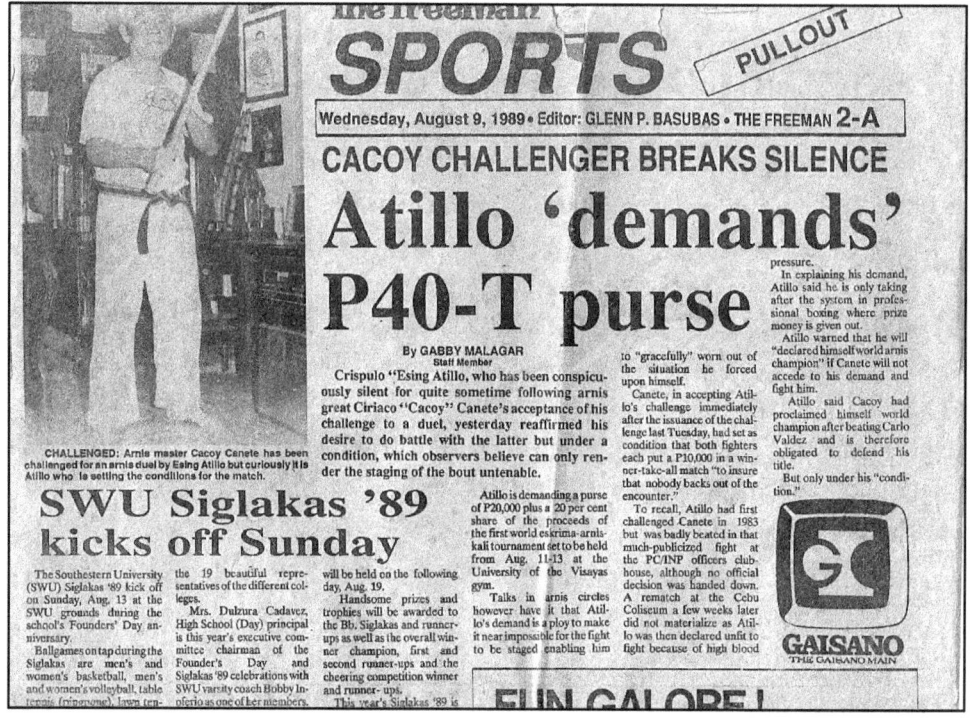

The Freeman newspaper, August 9, 1989, Cebu City, Philippines.

Following the event, I challenged Cacoy every chance I had. As always, he would never accept, and as the years passed by, the likelihood of he and I participating in a rematch diminished. He and I were both growing older, and he was no longer the athletic eskrimador he once was. I wasn't either. However, that didn't stop us from playfully challenging one another at every opportunity.

On August 12, 1989, at the First World Eskrima Kali Arnis Federation (WEKAF) Championships in Cebu City at the University of the Visayas, I again challenged Cacoy to a rematch. Cacoy and I were at the tournament to support our students; however, we found ourselves confronting one another once again. He and I immediately began negotiating terms of a rematch. Cacoy stated that he would fight as long as 10,000 pesos was deposited into a bank as a nonrefundable bond to ensure nobody backed out. I agreed and suggested that an additional 40,000 pesos and a percentage of ticket sales be awarded to the fight winner. Cacoy realized I was serious about fighting and immediately backed out. Jokingly I said, "Cacoy if you won't fight me, I will declare myself the World Champion just like you did at the First National Open Arnis Championships in 1979!"

Our final confrontation occurred several years later in 2000 at the World Eskrima Kali Arnis Federation (WEKAF) National Championships in Cebu City. I had been invited by the tournament organizers to demonstrate my style of Balintawak eskrima, and Cacoy was there as a special guest. This time it was Cacoy who confronted me asking for another fight. Cacoy was now 81 years old. The opportunity for us to participate in a fair and competitive rematch was long gone. Instead, I teasingly said, "Cacoy! We are both old men now! Maybe we should instead take part in a running contest!"

CHAPTER 8
A New Beginning

Shortly after my fight with Ciriaco "Cacoy" Cañete in 1983, Max M. Pallen Sr. flew to Cebu from the United States to continue his training in the Filipino Martial Arts and learn the Balintawak style of eskrima. Max had heard about my fight with Cacoy at the Philippine Constabulary Headquarters on September 17, 1983, and wanted to learn Balintawak eskrima.

Vicente "Inting" Atillo, Max M. Pallen Sr. and Crispulo "Ising" Atillo at the Atillo resident in Mambaling, Cebu City, Cebu Philippines.

Max was born and raised in Camarines Sur in the Bicol region of the Southern part of Luzon, Philippines, and studied eskrima as a child. After relocating to the United States, Max served as the United States and Regional Director of the World Eskrima Kali Arnis Federation (WEKAF) under Dionisio "Diony" Cañete. Max also became the US representative of Remy Presas, the founder of Modern Arnis, and created the Northern California Eskrima Kali Arnis Federation. Coincidentally, Remy Presas was a student of Arnulfo "Opong" Mongcal, a Balintawak eskrima student under my dad.

When Max arrived and checked into his hotel on Mactan Island[110], he asked one of the security guards where he could find an eskrimador to learn Balintawak eskrima from. Max had previously trained with the Cañetes at the

---

[110] Mactan Island is an island just off the eastern coast of Cebu. It is the most densely populated island in the Philippines and alleged location of the legendary battle between Portuguese explorer Ferdinand Magellan and Datu Lapu-Lapu.

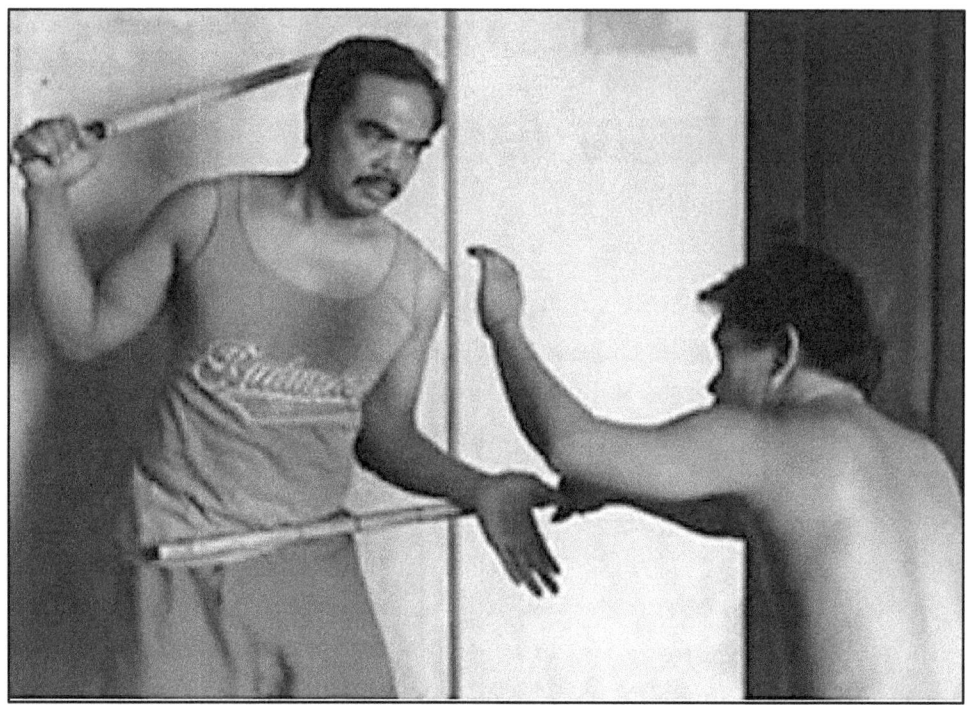
Crispulo "Ising" Atillo and Max M. Pallen Sr., Mambaling, Cebu City, Philippines, c. 1983.

Doce Pares Headquarters in Cebu City and advised the security guard that he also wanted to learn Balintawak eskrima. The security guard gave him my name because I had been training the military and security guards at the hotel and recommended that he train with me.

A few days later, Max arrived at my house in Mambaling and introduced himself to my dad and me. Max advised that he had heard our style of Balintawak eskrima was the best for actual fighting and asked if we would be willing to teach him. Max said he had been practicing other eskrima styles since he was a child but wanted to learn Balintawak eskrima. My dad and I performed a short demonstration for Max, who was immediately impressed. Max asked if he could check out of his hotel and stay at our home for his remaining time in Cebu to completely dedicate himself to training. In exchange, he offered to pay us for his lodging and training.

During his two-week stay, Max trained very hard every day, beginning early in the morning and often running late at night. When I was not available, my dad would take my place and train Max until I arrived. Max continued training for the remaining two weeks. He was in the Philippines and returned almost years to continue his training. Upon completion, He promised my dad he would help me come to America.

Passing of an Icon

On August 7, 1989, my dad fell ill. He was diagnosed at the Cebu City Medical Center with Pneumonitis[111] and Atherosclerosis[112]. His health began to steadily decline. His ultimate wish was to become an American citizen so that his children could move to the United States and pursue a better life outside of the Philippines.

As a member of the United States Armed Forces Far East (USAFFE) and famed Recognized Guerillas (RGs) under Lt. Col. James Cushing during World War II, my dad was entitled American citizenship under the March 27, 1942 Amendment to the Nationality Act of 1940. The act allowed Filipino veterans to become naturalized citizens of the United States to reward their service to the U. S. during the war. Unfortunately, after the surrender of the Empire of Japan on August 15, 1945[113], U.S. Congress passed the Rescission Act of 1945 on February 18, 1946, retroactively rescinding American citizenship and veterans' benefits to native Filipinos who served during the war.

Vicente "Inting" Atillo practicing with Crispulo "Ising" Atillo, Cebu City, Philippines, c. 1989.

---

[111] Pneumonitis is a type of pneumonia that is usually caused by a virus that leads to inflammation in the lungs.

[112] Atherosclerosis is a cardiovascular disease in which plaque builds up inside the arteries.

[113] The Emperor of Japan, Hirohito, announced the surrender of the Empire of Japan on August 15, 1945, however the Instruments of Surrender were not officially signed until September 2, 1945.

The act was a blow to patriots like my father. They sacrificed and loyally fought alongside U.S. service members throughout the war. These loyal Filipinos deserved recognition by the American government and earned naturalization and U.S. veterans' benefits under the earlier 1942 amendment. Nevertheless, U.S. citizenship was not what inspired and motivated veterans like my dad to fight against the Japanese. These men were dedicated and loyal Filipino patriots. They vowed to protect their families and defend their beloved island nation from those who threatened to deny the freedoms and democracy the country had fought so hard to establish. The promise of American citizenship instilled a sense of hope and a chance for a better life outside the war-ravaged Philippines for themselves and their children after the war.

Finally, on October 26, 1990, the U.S. Congress passed the Immigration Act of 1990. The updated act allowed for Filipinos' naturalization who served on active duty as members of the USAFFE, Philippine Army, Philippine Scouts, and recognized Guerilla Units between September 1, 1939, and December 31, 1946. This was an enormous victory for my dad and the Filipino veterans of World War II. They finally received recognition for their sacrifices and service to the United States during the war. Unfortunately, the law only applied to qualified veterans who could complete the United States Naturalization process. However, in 1993, the U.S. Congress passed a bill that permitted eligible World War II veterans to complete the Philippines' Naturalization process. Soon afterward, my dad started the process. By August 19, 1991, he had completed his background investigation. He received approval to participate in his final interview at the U.S. Embassy in Manila.

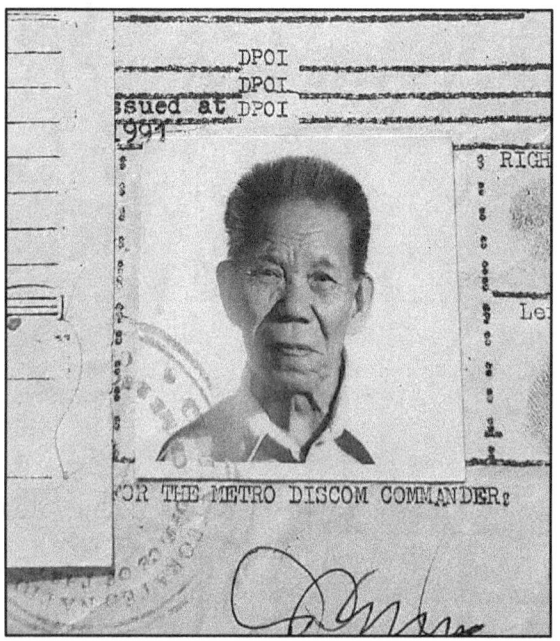

Vicente "Inting" Atillo Naturalization paperwork, Cebu City, Cebu, Philippines.

Crispulo "Ising" Atillo and Vicente "Inting" Atillo.

Sadly, my dad's health had declined to the point he was no longer able to travel and was eventually confined to his bed at home in Mambaling. Recognizing he was not going to be able to complete the process and immigrate to the United States as he had hoped, my dad said, "Son, I am dying. The spirit of Doring is in you, and you are the only one who has adopted Doring's style. Go to the United States and teach the art."

My dad was an early member of the Doce Pares Club and one of the Balintawak Self Defense Club co-founders in 1952 along with Venancio "Anciong" Bacon and Delfin Lopez. He had grown untrusting of his compatriots and had watched both clubs wither away from politics and petty jealousies. He had seen family bonds and close friendships destroyed due to childish rivalries and quarreling over who was a better eskrimador and the use of names like *Doce Pares* and *Balintawak*. My dad felt that Balintawak's continued use would only create more conflict amongst the quarreling factions of Balintawak eskrima that had been formed. He preferred to avoid politics and elected not to use the name toward the end of his life and encouraged me to avoid using it until I immigrated to the United States. Exhausted by the politics and feeling guarded, he said, "Son, don't share the complete art with the Filipinos and don't use the name *Balintawak* until you get to the United States."

It reminded me of the same politics and quarreling in 1952 when Anciong formed the Balintawak Self Defense Club and decided to no longer be a member of the Doce Pares Club, or use the *Doce Pares* name despite what others wanted. Although I was only 14 years old, I was present during many of the discussions. I was an original member of the Balintawak Self Defense Club, as was my dad and Delfin Lopez. The rivalry between the Balintawak Self Defense Club and the Doce Pares Club led to violence, divided political loyalties, and even murder. As an eskrimador, my dad eluded the politics and personal conflicts that plagued Cebuano eskrima for decades. He initially felt by discontinuing the use of the *Balintawak* name when he and I founded the New Arnis Confederation of the Visayas and Mindanao (NAC) in 1975, he would be able to put an end to much of the quarreling and remain a neutral friend with all eskrimadors. He later recognized that it wasn't possible. Instead, he chose to stay loyal to Anciong and the club he helped create by renaming our confederation the Balintawak World Arnis Association.

On June 8, 1993, my dad peacefully passed away at home from heart failure. He was a gentle person respected by everyone in the Cebuano eskrima community regardless of style or club affiliation. He ignored the never-ending politics and was one of the calming forces between the Doce Pares Club and Balintawak Self Defense Club. He was well-liked by everyone he encountered. He was a champion weight lifter in his youth, a wrestling champion of the Visayas and Mindanao, and a master eskrimador who had practiced and taught eskrima his entire life. He had personally known Lorenzo "Tatay Ensong" Saavedra and was a close friend and training partner of both Teodoro "Doring" Saavedra and Delfin Lopez. He had been a very close friend and training partner of Venancio "Anciong" Bacon and a respected friend and colleague of other legendary eskrimadors to include Timoteo "Timor" Maranga, Arnulfo Mongcal, and countless others.

My dad was also a quiet man who did not outwardly brag of his ability as an eskrimador. He also did not talk about his experiences as a member of the guerilla forces during World War II. He was a decorated war hero and Recognized Guerilla (RG) of both the Combat Company of the 85th Infantry and the famed "A" Company of 1st Battalion, 87th Infantry of the renowned Cebu Area Command (CAC) of the United States Armed Forces Far East (USAFFE) under the legendary Lt. Col. James Cushing. His passing was mourned throughout the Philippines and signified the end of an era of eskrimadors who lived during the early years of eskrima. He was preceded in death by his close *compadres* Teodoro "Doring" Saavedra and Delfin Lopez. Together they were the Tres Mosqueteros and unsung heroes of Balintawak eskrima.

Headstone of Vicente "Inting" Atillo (March 15, 1912 – June 8, 1993) at the Veteran's Cemetery, Cebu City, Philippines. The headstone reads, "PVT Vicente S. Atillo, II World War Veteran, "A" Company 1st BN 87th Infantry SN 132405, Weight Lifter, Wrestler & Arnis Grandmaster of the Balintawak World Arnis Assn."

Immigration to the United States

After my dad passed away, I continued teaching Balintawak eskrima at my school in Cebu City until late 2000. It was then that Max Pallen recontacted me invited me to come to the United States. Max asked me to conduct a demonstration at the upcoming World Eskrima Kali Arnis Federation (WEKAF) tournament in Northern California and teach a series of seminars on my style of Balintawak eskrima. Max offered to pay for my plane ticket and allowed me to stay at his residence during my time in the United States. With the help and sponsorship of Dionisio "Diony" Cañete, I obtained my visa, permitting me to travel to the United States. Despite the years of quarreling between his older brother and me, Diony has always been my friend and was instrumental in receiving my visa.

On January 26, 2001, I arrived in San Francisco, California, with only a stick and 100 pesos in my pocket. After being picked up by Max, I conducted a demonstration at the WEKAF tournament and taught several seminars in San Leandro, Los Angeles, Maine, Philadelphia, Texas, and New York and several locations throughout the Northern California area. I stayed at Max's house in San Leandro, California, and continued teaching Max privately during my stay. I was paid a small amount of money for the demonstration and seminars; however, I was not paid for teaching Max privately while I stayed at his residence. The only compensation I received was a new pair of shoes, which I appreciated. Because of the popularity of my demonstration at the tournament, there were a lot of people who wanted to learn my style of Balintawak eskrima. I also needed money so I could return to the Philippines. Unfortunately, Max didn't want me to stay and teach and encouraged me to return to Cebu. Instead, he said he would help me go back to the Philippines and offered to be my official representative in the United States. Although Max had been a loyal student of my dad and me in the Philippines since 1983 and followed through with his promise to help me come to America, I decline his offer. I instead chose to remain in the United States and make it on my own.

I then moved to Ramon Rubia's residence in Buena Park, CA. Ramon is an extremely talented eskrimador from Cebu and was the U.S. representative of the Doce Pares Club and the *San Miguel Eskrima*[114] style of Filemon "Momoy" Cañete. Ramon's wife, Eva, is Dionisio "Diony" Cañete's niece and is a very talented eskrimador herself. Ramon and his family were very generous, and he introduced me to several prominent martial artists in the area to include the world-famous Dan Inosanto. Shortly after Ramon introduced me to Dan, he and I taught a seminar together at Dan's school, the

---

[114] A style of eskrima founded by Filemon "Momoy" Cañete of the Doce Pares Club. The name San Miguel is taken from the archangel, Michael, the slayer of Satan. Like Vicente "Inting" Atillo, Momoy was a peacekeeper and member of the old-guard of Cebuano eskrima.

Crispulo "Ising" Atillo, Dan Inosanto, and Ramon Rubia at the Inosanto Academy of Martial Arts. (Photo courtesy of Ramon Rubia).

Inosanto Academy of Martial Arts. Dan was very impressed and asked if he could become my student. Dan has had numerous instructors of the Filipino Martial Arts throughout the years and has a very educated eye of what styles and systems are effective. He has been instrumental in exposing the Filipino Martial Arts to the public and has helped countless eskrimadors from the Philippines gain notoriety and teach their respective eskrima styles in the United States. Including members of both the Doce Pares Club and original Balintawak Self Defense Club. It was the first time since my arrival in the United States that I was being paid for teaching, and I was beginning to build a following of my own students as I had hoped.

A few months later, in June 2001, I moved from Ramon's residence in Buena Park, California, to Loma Linda, California, to live with my mom's first cousin, Lucy Abela, Venarendo Ylaya, and Alvin Ylaya. Living with my cousins in Loma Linda enabled me to be closer to Los Angeles and teach Dan and develop my own following of students in the United States. It was while living in Loma Linda that I met Dr. Jesse Devera. Dr. Devera was the Chief of Psychiatric Consultation and Liaison Services at Loma Linda University and later became the Regional Medical Director of the San Bernardino Country Department of Mental Health. He and his son Jordan had been taking *Pangamut* classes from Master Felix Pascua[115] in Loma Linda, California. Dr. Devera enjoyed training with Felix but told Jun Mendez, a mutual friend of ours, that he also wanted to learn Balintawak eskrima. Jun showed Dr. Devera a video of the seminar I conducted earlier in the year at the Inosanto Academy of Martial Arts and recommended he contact me for lessons.

---

[115] Master Felix Pascua, also known as Dong Meyong, is an extremely talented martial artist from Leyte, Philippines who teaches the Pangamut system of the Filipino Martial Arts in Loma Linda, California.

A short time later, Dr. Devera called and introduced himself. Dr. Devera asked if I would teach him and offered to pick me up every weekend to drive me to his residence in Beaumont, California, where we would practice. I continued teaching Dr. Devera until the end of 2002. At that time, he asked me to live at his residence and offered to build a small training facility in the back of his property where I could teach my own students. I was moving farther away from Dan Inosanto. Still, the move to Beaumont was going to enable me to create a base of students and have a permanent place to teach my style of Balintawak eskrima in the United States.

Not long after, I moved from my cousin's house in Loma Linda to Dr. Devera's residence in Beaumont and established the U.S. Headquarters of the Atillo Balintawak World Arnis-Eskrima Association. Just as my father had done years before when we renamed the Philippine Arnis Confederation to Balintawak World Arnis Association, I chose to remain loyal to my dad and the founding members of the Balintawak Self Defense Club. I renamed the association the Atillo Balintawak World Arnis-Eskrima Association after I immigrated to the United States.

Crispulo "Ising" Atillo and Dr. Jesse R. Devera, M.D., Beaumont, California.

Since there were other well-known masters of Balintawak eskrima already teaching in the United States, I recognized that I needed to differentiate my style of eskrima from the others. Masters such as Teodoro "Teddy" Buot and my friend Guillermo "Bobby" Taboada had already relocated to the United States and were teaching their own respective styles of Balintawak eskrima, as was Remegio "Remy" Presas, the founder of Modern Arnis. Remy had been a Balintawak eskrima student under Arnulfo Mongcal and briefly studied with Anciong before immigrating to the United

States. Mongcal had also been a student of my dad. Also, there were children of Balintawak eskrima masters in the Philippines who had immigrated to the United States. Many of them competent and qualified instructors. I instead chose to call my style Atillo Balintawak Eskrima - Saavedra Style, which I later changed to Original Saavedra Style. I chose the name to differentiate my style from other methods of Balintawak eskrima and identify it as the Balintawak lineage of my dad, and the style of Teodoro "Doring" Saavedra, which is the foundation of my eskrima.

Brothers of Balintawak

This is my history. The history of Atillo Balintawak Eskrima - Original Saavedra Style and my lineage of Balintawak eskrima. I purposely didn't include the history of other styles of Balintawak eskrima because they all have their own unique history that should be told separately by each respective master. Yet, they can all be called Balintawak eskrima and rightfully use *Balintawak* because they all originated from Venancio "Anciong" Bacon and the founding eskrimadors of the Balintawak Self Defense Club. Regardless of any modifications or improvements, each master has made.

Just as the Doce Pares Club was named after the Twelve Warrior Swordsman of King Charlemagne, who were united together as warrior-peers, the Balintawak Self Defense Club was named after Balintawak Street in Cebu City, Philippines. It symbolizes the unification of Filipino patriots against oppression and the beginning of the Philippine Revolution against the Spanish régime by the famed revolutionary society, the Katipunan, in 1896. Inherently, the same holds true for Balintawak eskrima today. Even though each style of Balintawak eskrima is uniquely different, we are all united together as brothers.

I have been blessed to have had the opportunity to immigrate to the United States and share my style of Balintawak eskrima with students from around the world. Today, I am still married to my beautiful wife, Beatrice "Betty" Enriquez Atillo, and we have four beautiful children, three boys, Rene, Nathaniel, and Crispulo Jr., and one girl, Marietta. I also have thirteen grandchildren, four granddaughters, and nine grandsons.

I have a hectic travel and seminar schedule and students who come from all around the world to learn from me at my training facility in Beaumont, California. Many of them are accomplished instructors and students of other eskrima styles, as well as members of law enforcement, the

U.S. Army Special Forces, and various U.S. Special Operations Forces and Counterterrorism personnel. I have produced an instructional DVD series on Atillo Balintawak Eskrima and have been featured in numerous national and international books and magazines. I was also honored to be conducted into the Martial Arts Hall of Fame as the Most Distinguished Grandmaster on May 15, 1999, and was awarded a Certificate of Appreciation from the Mayor of Los Angeles, California, in 2004. All of my success has been made possible due to the kindness of my friends and my beloved art of Atillo Balintawak Eskrima – Original Saavedra Style.

Mga kapatid ng Balintawak!

Angles of Attack

Most eskrima styles have twelve fundamental strikes that are taught to the student to teach the basic striking angles. There are only ten fundamental strikes used in Atillo Balintawak Eskrima. Crispulo "Ising" Atillo reduced the number of strikes to include only the most common angles encountered in actual combat.

Each strike is further broken down into *full-swings* and *half-swings*. A full- swing strike extends all the way outward and through the intended target with full power, whereas the half-swing stops mid-way to the target for purposes of training. They are used in training to teach proper body positioning and defensive counters to each of the ten fundamental striking angles.

Horizontal forehand strike to the opponent's left temple.

Horizontal backhand strike to the opponent's right temple.

Diagonal forehand strike to the opponent's left leg.

Diagonal backhand strike to the opponent's right leg.

Straight thrust to the center of the opponent's body.

Backhand thrust to the right side of the opponent's face.

Forehand thrust to the left side of the opponent's face.

Forehand uppercut to the left side of the opponent's body.

Backhand uppercut strike to the right side of the opponent's body.

Downward vertical strike to the top of the opponent's head.

4 Methods of Defense

In Atillo Balintawak Eskrima, there are four *methods* in every attack and four *methods* or sectors in every defense to counter an opponent's strike. Each of the ten striking angles can be defended against with the checking hand by touching and controlling the opponent's stick, not touching the opponent's stick, by touching and holding the opponent's hand or by direct defense of the opponent's attack. These four *methods* are further broken down by four defensive and offensive *sectors* from which each counter-attack occurs.

The following demonstrates the four basic *methods* or *sectors* used to defend against the opponent's attack by touching the stick for control.

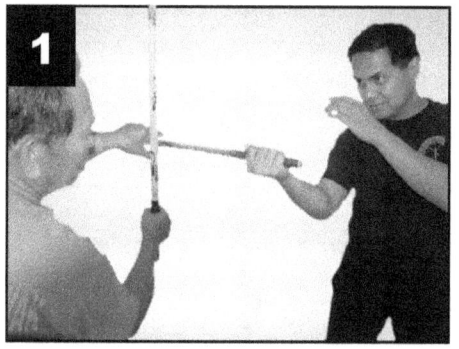

Left Side Sector No. 1. Left side of your stick above the opponent's stick.

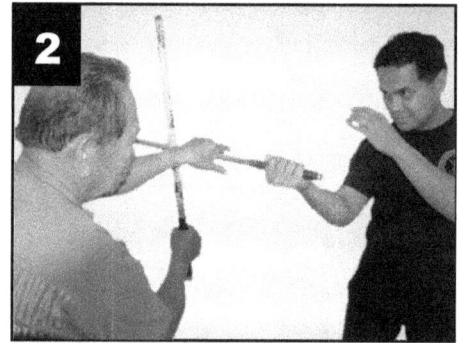

Left Side Sector No. 2. Right side of your stick above the opponent's stick.

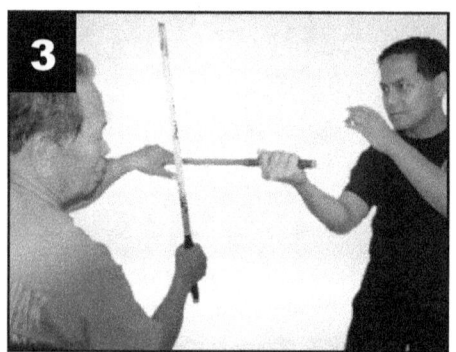

Left Side Sector No. 3. Left side of your stick above the opponent's stick (same as Left Side Sector No. 1).

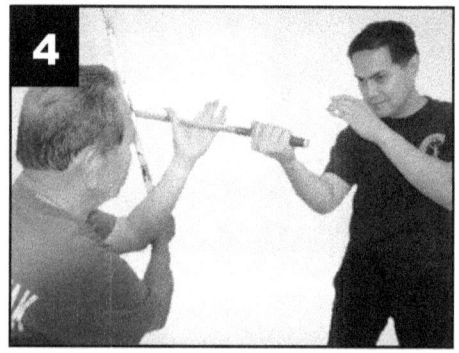

Left Side Sector No. 4. Right side of your stick beneath your opponent's stick.

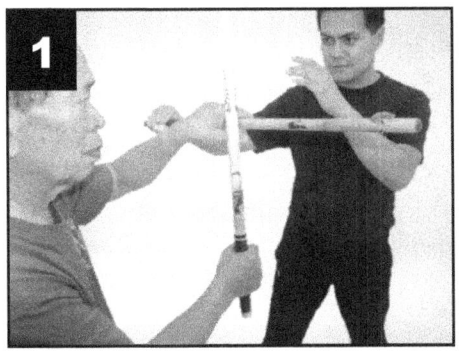

Right Side Sector No. 1. Left side of your stick above the opponent's stick.

Right Side Sector No. 2. Right side of your stick above the opponent's stick.

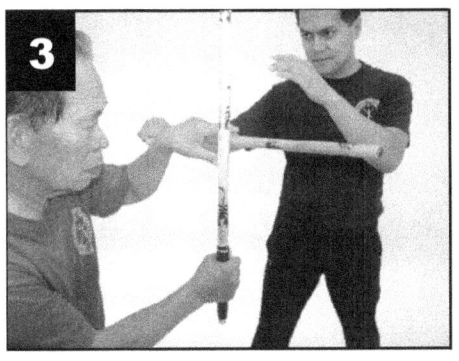

Right Side Sector No. 3. Left side of your stick above the opponent's stick (same as Right Side Sector No. 1).

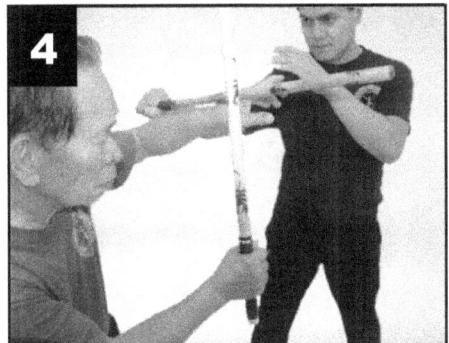

Right Side Sector No. 4. Left side of your stick beneath your opponent's stick.

4 Methods of Attack

Just as there are four *methods* in every defense, there are four *methods* or *sectors* in every attack when the opponent blocks your attack with his stick or hand. The following demonstrates the four basic *methods* or *sectors* used to control and counter the opponent's stick when attacking.

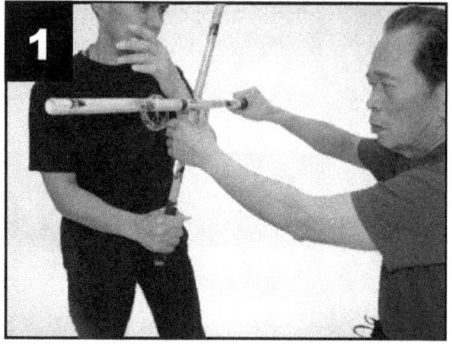

Left Side Sector No. 1. Left side of your opponent's stick below your stick.

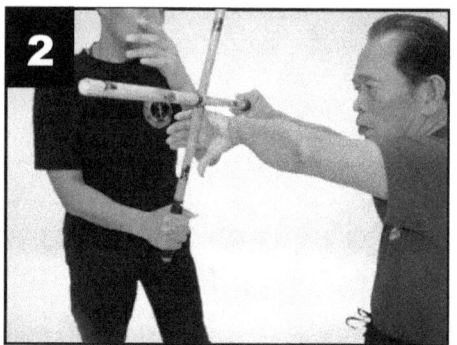

Left Side Sector No. 2. Right side of your opponent's stick below your stick.

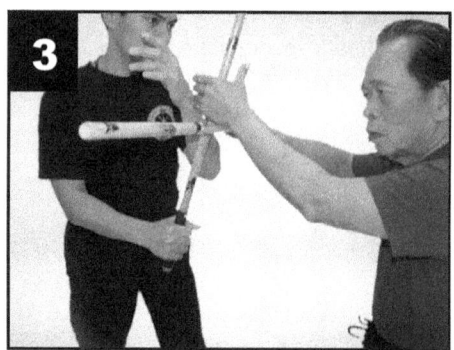

Left Side Sector No. 3. Left side of your opponent's stick above your stick.

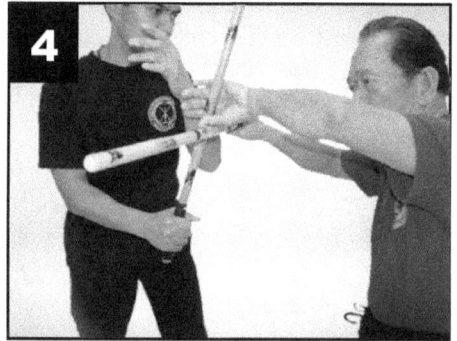

Left Side Sector No. 4. Right side of opponent's stick above your stick.

Right Side Sector No. 1. Right side of your opponent's stick above your stick.

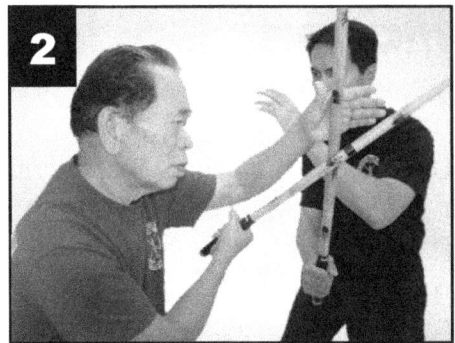

Right Side Sector No. 2. Left side of your opponent's stick above your stick.

Right Side Sector No. 3. Right side of your opponent's stick below your stick.

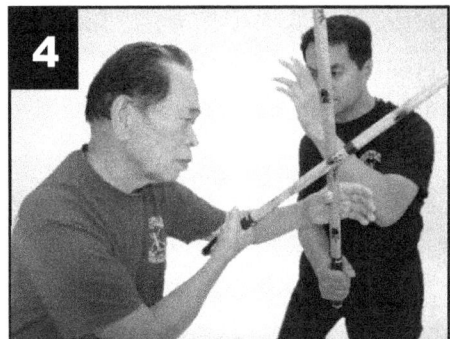

Right Side Sector No. 4. Left side of opponent's stick below your stick.

Mother Spar Drill with Stick

Once the fundamental and intermediate techniques of Atillo Balintawak Eskrima are mastered, the student then progresses through the advanced or *College Level* curriculum. The advanced level curriculum of Atillo Balintawak Eskrima is comprised of over thirty-six drills, each of which teaches the student a separate technique or counter unique to Atillo Balintawak Eskrima. In addition, numerous options are introduced within each drill, which can be performed with a *single stick*, *single dagger*, *bolo*, or *empty hands*.

The first drill taught to the student as part of the advanced level curriculum is the *Mother Spar*. All other drills at the *College Level* are variations of this drill, which serves as the most important drill taught to the student.

GGM Atillo and Master Derrick Dalan face each other.

GGM Atillo strikes Derrick with an Angle No. 1 that Derrick defends with a vertical block.

Derrick defends and controls GGM Atillo's stick by using Method No. 1 in Sector No. 1.

Derrick controls GGM Atillo's stick and executes a backhand Angle No. 2 strike.

GGM Atillo defends Derrick's backhand Angle No. 2 with his checking hand by parrying it downward.

GGM Atillo then counters with an Angle No. 1 strike which Derrick defends with his checking hand.

GGM Atillo counters Derrick's grab by pulling and releasing his stick from Derrick's checking hand.

GGM Atillo then counters with a backhand Angle No. 2 strike which Derrick counters with a vertical block.

Derrick controls GGM Atillo's stick by using Method No. 1 in Sector No. 1.

Derrick continues controlling GGM Atillo's stick and prepares to throw an Angle No. 1 strike.

Derrick executes an Angle No. 1 strike and GGM Atillo counters with a vertical block.

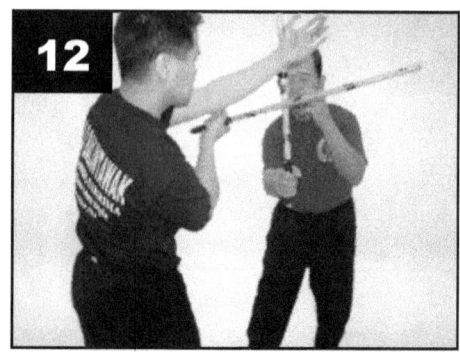

Derrick counters by clearing GGM Atillo's stick through Sector No. 1.

Derrick then counters with an Angle No. 1 strike.

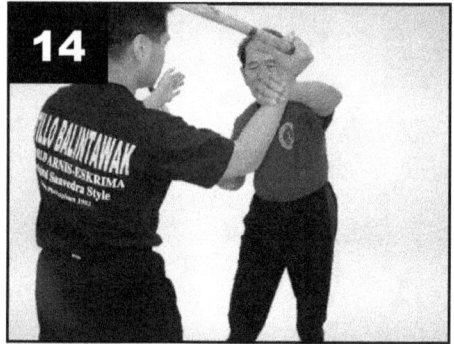

GGM Atillo counters and stops the strike with his checking hand.

Derrick counters GGM Atillo's grab by clearing his checking hand.

GGM Atillo counters with an Angle No. 1 to repeat the Mother Spar drill.

Mother Spar Empty Hands

All of the advanced or *College Level* curriculum drills can be performed with the empty hands by incorporating various strikes, defensive and counter motions, as well as trips, off-sets, and throws. The empty hand drills of Atillo Balintawak Eskrima have been described as being similar to the chi sao, or *sticky hands* exercise of Wing Chun gung fu. However, the empty hand drills of Atillo Balintawak Eskrima are uniquely Filipino.

GGM Atillo attacks Derrick with an Angle No. 1 empty hand strike which Derrick blocks with his right forearm.

Derrick counters with an Angle No. 2 empty hand strike which GGM Atillo blocks with his checking hand.

GGM Atillo controls Derrick's right hand and counters with an Angle No. 1 empty hand strike which Derrick blocks.

GGM Atillo counters Derrick's grab by pulling and releasing Derrick's checking hand.

GGM Atillo then counters with a backhand Angle No. 2 empty hand strike which Derrick counters with a vertical block.

Derrick then counters with a forehand Angle No. 1 empty hand strike which GGM Atillo blocks with his checking hand.

Derrick counters GGM Atillo's grab by clearing his checking hand.

GGM Atillo counters with an Angle No. 1 to repeat the empty hand Mother Spar drill.

Disarm Techniques

Disarming techniques are an essential part of Atillo Balintawak Eskrima. A practitioner should be able to disarm the opponent at any time from any position. Specific drills are used to develop a student's sensitivity and ability to feel the correct position and choose the appropriate disarm. Atillo Balintawak Eskrima specializes in disarming techniques, and the following is a collection of sample disarms techniques.

Derrick attacks GGM Atillo with an Angle No. 1 which GGM Atillo counters with a vertical block.

GGM Atillo grabs Derrick's stick using Method No. 1.

GGM Atillo places his stick behind the punyo of Derrick's stick.

GGM Atillo secures Derrick's stick under his arm and grabs his stick. GGM Atillo then pulls backward and disarms Derrick's stick.

Derrick attacks GGM Atillo with an Angle No. 1 which GGM Atillo counters with a vertical block.

GGM Atillo grabs Derrick's stick using Method No. 1.

GGM Atillo drops the point of his stick downward behind Derrick's stick.

GGM Atillo places Derrick's stick under his arm and grabs the end of his own stick to create leverage.

Holding on to his stick, GGM Atillo rotates his body clockwise and disarms Derrick's stick.

Derrick attacks GGM Atillo with an Angle No. 1 which GGM Atillo counters with a vertical block.

GGM Atillo grabs Derrick's stick using Method No. 2.

GGM Atillo wraps his forearm underneath Derrick's stick.

GGM Atillo pulls his forearm backward disarming the stick.

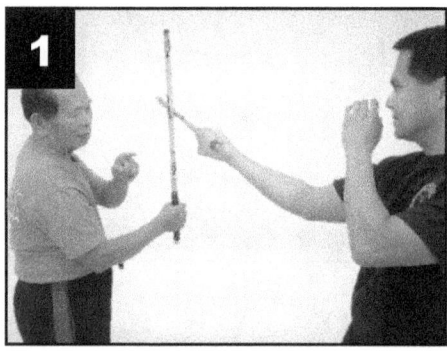

Derrick attacks GGM Atillo with an Angle No. 1 which GGM Atillo counters with a vertical block.

GGM Atillo defends with Method No. 2 and places his stick behind Derrick's.

GGM Atillo used Derrick's stick as a lever and continues pushing Derrick's stick downward.

GGM Atillo continues twisting and disarms Derrick's stick.

Derrick attacks GGM Atillo with an Angle No. 1 which GGM Atillo counters with a vertical block.

GGM Atillo counters with Method No. 2 to control Derrick's stick.

GGM Atillo pulls Derrick's stick close to his and traps it against his body.

GGM Atillo turns his body inward and uses his shoulder to disarm Derrick's stick.

Derrick attacks GGM Atillo with an Angle No. 1 which GGM Atillo counters with an inside vertical block.

GGM Atillo counters with Method No. 2 to control Derrick's stick.

GGM Atillo then places the punyo of his stick behind Derrick's wrist.

GGM Atillo continues rotating his body to the right and disarms Derrick's stick.

Derrick attacks GGM Atillo with an Angle No. 2 strike which GGM Atillo counters with an outside vertical block.

GGM Atillo grabs Derrick's stick using Method No. 1.

GGM Atillo secures Derrick's stick and pins it between his arm and body for control.

GGM Atillo then disarms Derrick's stick by striking downward with his left hand against Derrick's wrist while rotating his hips clockwise.

Derrick attacks GGM Atillo with an Angle No. 2 strike which GGM Atillo counters with an outside vertical block.

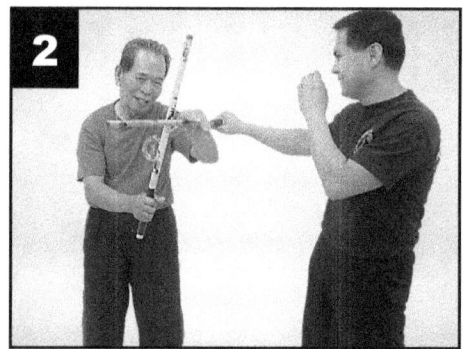

GGM Atillo counters with Method No. 1 and controls Derrick's stick.

GGM Atillo rotates his stick clockwise and underneath Derrick's stick.

GGM Atillo continues rotating his stick and completes the disarm.

Mother Spar with Knife

The advanced or *College Level* drills of Atillo Balintawak Eskrima performed with a knife or bolo generally follow the same pattern as those with a stick. However, small modifications are made to take into consideration the cutting edge of a bladed weapon. These drills are performed both *knife vs knife* and *knife vs empty hands*.

The knife tactics and techniques of Atillo Balintawak Eskrima are unique to the Teodoro "Doring" Saavedra style of eskrima and were taught directly to GGM Atillo by Delfin Lopez.

GGM Atillo attacks Derrick with an Angle No. 1 reverse thrust. Derrick defends and counters with an Angle No. 2 reverse thrust.

GGM Atillo counters with his checking hand and counters with an Angle No. 1 reverse thrust. Derrick counters with his checking hand.

GGM Atillo counters Derrick's grab by lifting and releasing Derrick's checking hand.

GGM Atillo then counters with a backhand Angle No. 2 thrust which Derrick counters with a vertical block.

Derrick counters using Method No. 1 and control GGM Atillo's arm through Sector No. 1.

Derrick then counters with an Angle No. 2 reverse thrust which GGM counters with his checking hand.

Derrick clears GGM Atillo's checking hand as GGM Atillo counters with an Angle No. 1

Derrick blocks and counters GGM Atillo.

Derrick continues the drill.

Knife Disarms

Disarming a knife is inherently dangerous and difficult to successfully perform under real conditions. However, it is important to have a fundamental knowledge of disarms against a knife. The knife tactics and techniques of Atillo Balintawak Eskrima are unique to the Teodoro "Doring" Saavedra style of eskrima and were taught directly to GGM Atillo by Delfin Lopez.

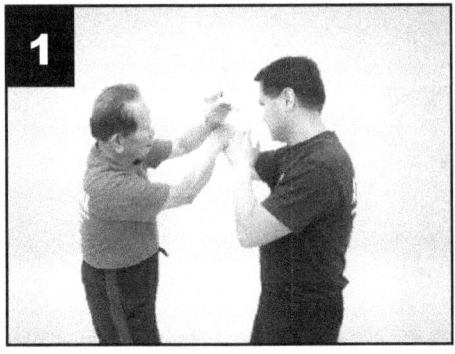

GGM Atillo momentarily blocks Derrick's Angle No. 1 reverse thrust with an outward empty hand block.

GGM Atillo immediately redirects the motion of the blade and applies pressure to Derrick's wrist.

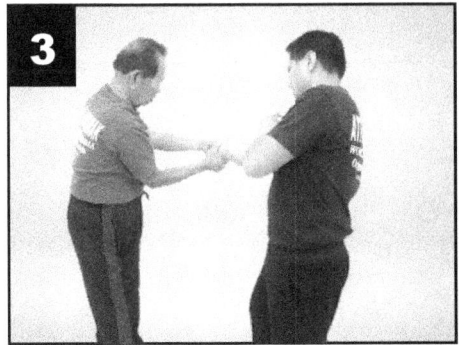

GGM Atillo then secures a wrist lock and disarms the knife.

GGM Atillo momentarily blocks Derrick's backhand Angle No. 2 reverse thrust with an outside empty hand block.

GGM Atillo redirects the knife and maintains control of Derrick's wrist.

Maintaining control of the knife, GGM Atillo secures a grip around Derrick's fingers holding the knife.

GGM Atillo then redirects Derrick's hand and disarms the knife.

Derrick attempts to stab GGM Atillo with an Angle No. 5 thrust to his midsection. GGM Atillo side-steps and redirects the thrust.

GGM Atillo maintains control of Derrick's wrist and secures the knife with both hands.

Maintaining control of the knife, GGM Atillo secures a grip around Derrick's fingers holding the knife.

GGM Atillo continues the motion and redirects the knife.

GGM Atillo then secures a wrist lock and disarms the knife.

Defense Against Hand Grabs

Master Derrick Dalan grabs GGM Atillo's stick hand with this checking hand.

GGM Atillo wedges the outside of Derrick's wrist and prepares to block his attack.

GGM Atillo maintains control of Derrick's wrist and prepares for a counter strike.

GGM Atillo twists his stick to release Derrick's grab.

Derrick grabs GGM Atillo's stick hand and prepares to strike GGM Atillo with an Angle No. 1 strike.

GGM Atillo places his checking hand behind his stick to defend against Derrick's strike.

GGM Atillo defends against Derrick's strike with a vertical block.

GGM Atillo grabs Derrick's stick and places his stick on the side of Derrick's neck.

GGM Atillo momentarily locks Derrick's stick in place.

GGM Atillo disarms Derrick's stick using a vine disarm.

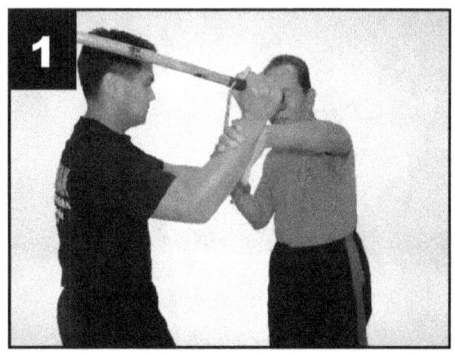

Derrick grabs GGM Atillo's stick and executes an Angle No. 1 strike. GGM Atillo blocks with his checking hand.

GGM Atillo wraps his hand around Derrick's stick to trap his stick in place.

GGM Atillo secures Derrick's wrist and releases his stick.

GGM Atillo then disarms Derrick's stick by striking downward across his arms.

Family Tree

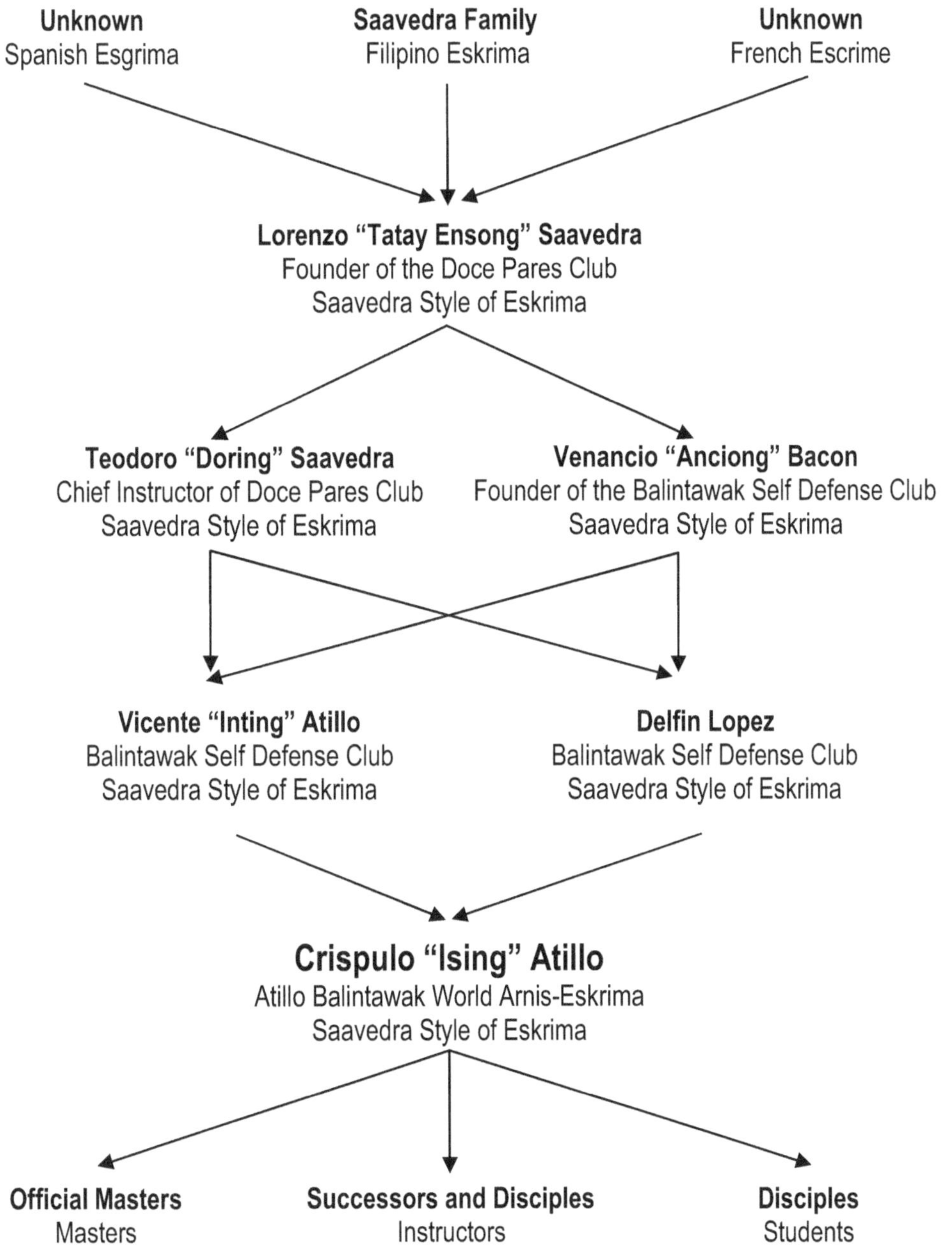

Unknown
Spanish Esgrima

Saavedra Family
Filipino Eskrima

Unknown
French Escrime

↓

Lorenzo "Tatay Ensong" Saavedra
Founder of the Doce Pares Club
Saavedra Style of Eskrima

↓

Teodoro "Doring" Saavedra
Chief Instructor of Doce Pares Club
Saavedra Style of Eskrima

Venancio "Anciong" Bacon
Founder of the Balintawak Self Defense Club
Saavedra Style of Eskrima

↓

Vicente "Inting" Atillo
Balintawak Self Defense Club
Saavedra Style of Eskrima

Delfin Lopez
Balintawak Self Defense Club
Saavedra Style of Eskrima

↓

Crispulo "Ising" Atillo
Atillo Balintawak World Arnis-Eskrima
Saavedra Style of Eskrima

↓

Official Masters
Masters

Successors and Disciples
Instructors

Disciples
Students

The lineage diagram illustrated above is specific to Atillo Balintawak Eskrima only and is not a complete representation of the entire Balintawak eskrima family tree.

Atillo World Balintawak

The following are the **"official"** *Masters*, *Successor & Disciples*, and *Disciples* of Atillo Balintawak World Arnis-Eskrima, Original Saavedra Style as of **August 5, 2018**. Please note, any names not listed below who have received Masters before this date are null and void due to not having the complete system, inactivity, and/or questionable character.

MASTER: "... is awarded the honor and rank of Master having satisfactorily trained, completed, and fulfilled the requirements for the rank of Master through the personalized guidance and keen instructions of GGM Atillo as governed by the standards of the Atillo Balintawak World Arnis-Eskrima, Original Saavedra Style".

Darrick Dalan   **Dennis Dalan**

SUCCESSOR AND DISCIPLE: "... is chosen and awarded the honor of Successor and Disciple having acquired exceptional knowledge, skills and ability of an Eskrima practitioner through the personalized guidance and keen instructions of GGM Atillo as governed by the standards of the Atillo Balintawak World Arnis-Eskrima. Your responsibility is to continue learning, to represent, and to propagate the art of Atillo Balintawak Original Saavedra Style of Eskrima."

Alberto Figueroa

Alfredo Parayno

Allan Sargan

Andrew Do

Andy Koh

Bernard "Butch" Sepulveda

Che Navidad

Chris Callahan

Daniel Lanero

Dieter Roser

Don Saavedra Briones

Doug Pierre

Edgar Gabriel

Eric Sutz

Felix Conde Sr.

Felix Conde Jr.

Geyo Esmas

Glen Boodry

Greg Sepulveda

Harley Elmore

Ike Sepulveda

Dr. Jesse Devera

Joe Medrano

Joel Clark

John Piastuck

Jordan Devera

Juerg Ziegler

Luiz Bonfá

Marc Halleck

Marco Lebrera

Nino Pilla

Ramir General

Ramon Rubia

Ray Jurado

Ron Goldstein

Salem Assli

Sam Halim   **Tom Bolden**

The above list of *Successors and Disciples* is not all-inclusive and reflects only those who are active members of the ABWAE. It does not include those who are in inactive status or who did not submit photographs in the correct format for publication.

Not pictured above include:

Alex Mancao	Lavonne Martin
Alvin Ylaya	Michael Bates
Dennis Flores	Neil Cauliffe
Helena Cauliffe	Pete Hsu
JD Lopez	Peter Vargas
Katherine Straus	Quinten Egson

DISCIPLE: "... is chosen and awarded the honor of *Disciple* having acquired exceptional knowledge, skills, and ability of an Eskrima practitioner through the personalized guidance and keen instructions of GGM Atillo as governed by the standards of the Atillo Balintawak World Arnis Eskrima. Your responsibility is to continue learning, to represent, and to propagate the art of Atillo Balintawak Original Saavedra Style of Eskrima."

Aimee Zapata

Amir Deleon

Anthony Gaudio

Armen Zargarian

Brett Ryan

Denard Harris

Eddie Wong

Edwin Williams

George Dalan

Harvey Sevilleno   **Jason Lin**   **Jeffrey McCready**

Jesus Lisojo   **Johnny Bosch**   **John Fasanaro**

Jose Conde   **Juan Mendez**   **Krystal Conde**

Orben Pastrana

Pete Italiano

Ron Balisi

Shervin Elbeig

Tim Becherer

Tim Ferris

Zena Aquino

Zhixiang Ma

The above list of *Disciples* is not all-inclusive and reflects only those who are active members of the ABWAE. It does not include those who are in inactive status or who did not submit photographs in the correct format for publication.

Not pictured above include:

Bong Nibres	Larry Atil
Brando Haworth	Michael Martinez
Christopher Bruce	Ric Lin
David Murphy	Rovalito Piodos
Jeff Morris	Sam Libby
Ken Lehrer	Peter Hsu

Timeline of Events

1852   Lorenzo "Tatay Ensong" Saavedra is born in the city of Carcar, Cebu, Philippines

Old railway station, Carcar, Cebu, Philippines.

January 20, 1872   Approximately 200 military personnel and general laborers stationed at the Fort San Felipe revolted against the Spanish government.

Fort San Felipe, Cavite, Luzon, Philippines.

January 27, 1872   Governor-General Rafael Izquierdo approves the death sentences of forty-one mutineers. On February 6, 1872, eleven more are sentenced to death.

February 17, 1872	Three Filipino priests are implicated in the revolt and executed. Their execution gives birth to the Reform Movement amongst Filipinos living in Spain.

GOMBURZA Memorial in Rizal Park, Manila, Philippines.

1872	Spanish authorities begin to arrest, imprison, exile, and execute anyone suspected of organizing or supporting anti-government activities or who even speak out against the Spanish government.
	Lorenzo "Tatay Ensong" Saavedra is accused of being a rebel and is incarcerated in the *Cárcel de Cebú*. While incarcerated, Tatay Ensong befriends a French prisoner who begins teaching him the French method of traditional European sword fighting, or *escrime*.
July 7, 1892	The Philippine revolutionary society, the Katipunan, is officially established in Tondo, Manila, Philippines, by Filipino patriots, Andres Bonifacio, Teodoro Plata, and Ladislao Diwa.
August 23, 1896	The Philippine Revolution begins with the *Cry of Balintawak*. The leaders of the Katipunan organize themselves into a revolutionary government, the

Haring Bayang Katagalugan, and openly declare a nationwide armed revolution against the Kingdom of Spain.

April 21, 1898   Spain severs diplomatic relations with the United States.

April 25, 1898   The United States declares war on Spain, resulting in the Spanish-American War.

August 12, 1898   Hostilities are halted between the United States and Spain with the signing of a Protocol of Peace.

Filipino soldiers during the Spanish-American War.

August 13, 1898   The Philippine Revolution ends.

December 10, 1898   The Spanish-American War ends with a decisive victory by the United States and the signing of the Treaty of Paris. Spain relinquishes all claim of sovereignty over Puerto Rico, Guam, and the Philippines to the United States.

February 4, 1899 The Philippine - American War begins as fighting erupts between forces of the United States and those of the Philippine Republic.

July 2, 1902 The Philippine - American War ends.

July 4, 1902 Proclamation 483 is signed by United States President Theodore Roosevelt, granting full pardons and amnesty to all persons in the Philippines who participated in or supported insurrection activities against Spain and the United States between August 1896 until the cession of the Philippines by the United States in 1898.

October 24, 1911 Teodoro "Doring" Saavedra is born in the San Nicolas District of Cebu City, Philippines.

Cebu Railroad Station, 1910.

March 15, 1912 Vicente "Inting" Atillo is born in Mambaling, Cebu City, Philippines.

October 15, 1912 Venancio "Anciong" Bacon is born in the Carcar district of Cebu City, Philippines.

August 29, 1916 The Jones Act is enacted by the United States Congress formally declaring the intention of the United States to grant independence to the Philippines as soon as a stable government can be established.

December 24, 1917 Delfin Lopez is born in Cebu City, Philippines.

August 14, 1920 Lorenzo "Tatay Ensong" Saavedra is released from prison and organizes the Labangon Fencing Club, the first formal eskrima club in the Philippines.

Cebu Provincial Jail, 1915.

August 14, 1930 The Labangon Fencing Club is disbanded due to political infighting and a dispute over missing money donated to the club.

January 11, 1932 Lorenzo "Tatay Ensong" Saavedra founds the Doce Pares Club along with Teodoro "Doring" Saavedra and Eulogio "Yoling" Cañete.

January 21, 1932 Membership in the Doce Pares Club increases, and a formal election of officers is conducted.

September 1933	Teodoro "Doring" Saavedra fights and defeats Pablo "Ambong" Alicante in Argao, Philippines.
1937	Teodoro "Doring" Saavedra fights and defeats Pedrito "Pedring" Romo in Pasil, Cebu City, Philippines.
1938	Teodoro "Doring" Saavedra fights and defeats Roman "Oman" Ladaño in Mambaling, Cebu City, Philippines.
June 10, 1938	Crispulo "Ising" Atillo is born in the Mambaling district of Cebu City, Philippines.
July 26, 1941	The United States Armed Forces in the Far East (USAFFE) is formed in response to the escalating conflict throughout Asia and establishes its headquarters in Manila, Philippines. General Douglas MacArthur is recalled from retirement by the U.S. War Department and named commanding officer of USAFFE.

General Douglas MacArthur, Commanding Officer of USAFFE (January 26, 1880 - April 5, 1964)

December 8, 1941	The Empire of Japan declares war on the United States and attacks the U.S. Naval Station at Pearl Harbor, Hawaii (December 7, 1941, U.S. time).

Japanese forces attack the Philippine Islands. In a series of airstrikes, Japanese aircraft are able to destroy over half of the Far East Air Force (FEAF) aircraft at Clark and Iba Fields in Northern Luzon. At the same time, a Japanese landing force makes an unopposed landing at Batan Island in the Luzon Strait and seizes control of the airstrip on the island.

Honolulu Star Bulletin, December 7, 1941.

December 8, 1941   The United States declares war on the Empire of Japan in response to the surprise attack on Pearl Harbor, Hawaii.

December 10, 1941   Japanese forces conduct airstrikes against the Philippines and inflict heavy damage to Del Carmen Field, Clark Field, Nichols Field, Nielson Field, and the US Naval Facility at the Cavite Naval Yard. Japanese forces conduct landings at Camiguin Island, Vigan, Aparri and Gonzaga along the northern coast of Luzon.

December 12, 1941   A landing force of 2,500 Japanese soldiers land at Legazpi on the southernmost point of Luzon.

December 19, 1941 Admiral Thomas C. Hart, Commander of the US Asiatic Fleet, withdrawals the majority of the US Asiatic Fleet from the Philippines.

December 19, 1941 Japanese forces conduct an amphibious landing on Mindanao near Davao City, Philippines.

December 22, 1941 The main Japanese invasion of the Philippine Islands begins. Japanese troops under the command of General Masaharu Homma land at three separate locations along the Lingayen Gulf on the west coast of Luzon, northwest of Manila

Japanese soldiers invade the Philippines, 1941.

December 24, 1941 Japanese forces seized the initiative and land a second wave of Japanese troops at Lemon Bay, in the southern Luzon and begin their advance inland toward Manila. General Douglas MacArthur orders the remaining American and Filipino forces on Luzon to withdraw to defensive positions on the Bataan Peninsula.

December 25, 1941 General Douglas MacArthur, together with Manuel L. Quezon, the President of the Commonwealth of the Philippines, evacuate

Manila and relocate to the heavily fortified island of Corregidor.

December 26, 1941 General Douglas MacArthur abandons all defensive efforts and issues a proclamation officially declaring Manila an "Open City."

December 30, 1941 Japanese forces close off the Bataan Peninsula.

Japanese tanks advance toward the Bataan Peninsula.

March 11, 1942 U.S. President Franklin D. Roosevelt orders General Douglas MacArthur to relocate to Australia.

March 12, 1942 General Douglas MacArthur delegates command of all American and Filipino forces to General Jonathan "Skinny" Wainwright. MacArthur, his family, and several USAFFE staff officers evacuate Corregidor and escape to Mindanao on four US Navy PT boats commanded by Lieutenant Commander John D. Bulkeley.

March 17, 1942 General MacArthur arrives at Batchelor Field in the Northern Territory of Australia. He is then flown to Alice Springs and is transported through the Australian outback by passenger train to Adelaide.

March 20, 1942 General Douglas MacArthur makes a statement to the press from Terowie, South Australia, promises his men and the people of the Philippines, "I shall return."

March 21, 1942 Following his arrival in Adelaide, MacArthur relocates to Melbourne, Australia.

March 27, 1942 U.S. Congress passes the Second War Powers Act (Public Law 77–507; 56 Stat. 176) and promises a simplified naturalization process and American citizenship for Filipinos who served in the United States Armed Forces as well as full veteran's benefits. The Act included Filipinos who served in the Commonwealth Army of the Philippines, Philippine Scouts, and later members of the Recognized Guerillas. The act allows Filipino Veterans to become naturalized citizens of the United States as a reward for their service in World War II.

March 28, 1942 Japanese forces launch a wave of air and artillery attacks on the Allied troops on the Bataan Peninsula who are severely weakened by malnutrition, sickness and prolonged fighting.

April 3, 1942 Japanese forces launch their final assault on the Allied troops defending the Bataan Peninsula.

April 9, 1942 Major General Edward P. King surrenders the Bataan Peninsula to Major General Kameichiro Nagano and the Empire of Japan. The surrender leads to the transfer of all American and Filipino prisoners of war to a POW camp in Tarlac that later becomes known as the "Bataan Death March."

Bataan Death March, 1942.

April 10, 1942	Japanese forces invade the island of Cebu, Philippines. Vicente "Inting" Atillo, Delfin Lopez and Teodoro "Doring" Saavedra escape Cebu City into the interior mountains to join the fledgling guerilla resistance movement. Venancio "Anciong" Bacon escapes and flees to Carcar. Crispulo "Ising" Atillo takes shelter in Mambaling with his family.
April 18, 1942	Allied Forces establish General Headquarters (GHQ), South West Pacific Area (SWPA) Command in Melbourne, Australia. Gen. Douglas MacArthur becomes the Commander-in-Chief.
May 5, 1942	Japanese forces land on the island of Corregidor with additional reinforcements landing throughout the night.
May 6, 1942	Brigadier General Lewis C. Beebe, Assistant Chief of Staff of U.S. Forces under General Jonathan Wainwright, broadcasts over the "Voice of Freedom" radio an offer to surrender Corregidor. General Jonathan "Skinny" Wainwright formally

surrenders the Philippine Islands to the Empire of Japan.

Lieutenant General Jonathan M. Wainwright surrender negotiations with Lieutenant General Masaharu Homma, 1942.

May 10, 1942 Brigadier General William Sharp, Commander of the Visayan-Mindanao Force, surrenders to Japanese forces.

May 16, 1942 Colonel Bradford Chynoweth, Commander of the 61st Philippine Division, surrenders the island of Cebu to Japanese forces. Captain James M. Cushing of the USAFEE Corps of Engineers refuses to surrender and escapes into the mountainous interior of Cebu, Philippines.

American and Filipino troops who evaded capture begin organizing small guerilla resistance groups throughout the Philippine Islands.

July 1942 Hilario "Dodong" Abellana, the governor of Cebu, collaborates with Lieutenant José Macabuhay and secretly assembles a small group of patriots from Mambaling to include Vicente "Inting" Atillo, Teodoro "Doring" Saavedra, and Delfin Lopez. The newly formed guerilla group immediately

	begins conducting reconnaissance missions and harassment raids against Japanese forces.
August 1942	Captain James M. Cushing agrees to unite the various guerilla groups operating independently throughout the island of Cebu into a single fighting force. Cushing and Harry Fenton consolidate the guerilla groups into a single joint command called the Cebu Area Command (CAC)
August 1942	Vicente "Inting" Atillo, Teodoro "Doring" Saavedra and Delfin Lopez conduct a demonstration of boxing, wrestling, and eskrima for Captain James M. Cushing.
August 15, 1942	Vicente "Inting" Atillo, Teodoro "Doring" Saavedra and Delfin are formally inducted into the United States Armed Forces Far East (USAFFE) and assigned to the Combat Company of the 85th Infantry Regiment commanded by Lt. Rogaciano "Popoy" C. Espiritu and Lt. Macabuhay.
	Vicente "Inting" Atillo, Teodoro "Doring" Saavedra and Delfin Lopez begin teaching eskrima and hand-to-hand combat to their fellow guerilla fighters and form the Doce Pares of Tabunan.
August 30, 1942	Vicente "Inting" Atillo, Teodoro "Doring" Saavedra, Delfin Lopez, and the newly formed Cebuano guerilla unit received orders to kill a Filipino undercover working for the Japanese named Mariano T. Jaucian.
October 24, 1942	The Cebuano guerillas engaged in their first major battle against Japanese forces in the Battle of Babag.

December 28, 1942   The Cebuano guerillas conduct a major train assault at Inayawan Crossing.

January 17, 1943   Governor Hilario "Dodong" Abellana escapes to the neighboring island of Bohol.

February 1943   Japanese forces erect a fence around Cebu City to control access into the city and protect themselves from attacks by the Cebuano guerillas.

1943   The Cebuano guerillas participate in a major engagement against Japanese troops that later becomes known as the Farmhouse Ambush.

September 15, 1943   Harry Fenton is tried and executed by the Cebuano guerillas under his command at the CAC GHQ at Tabunan for a series of reckless and injudicious actions.

Filipino guerillas with captured Japanese soldiers.

October 1943   Teodoro "Doring" Saavedra is captured in Cebu City, Philippines and transported to the Basak Elementary School, where he is tortured and killed by the Japanese Kempeitai.

Japanese troops conduct a raid on Cebu City, Philippines, to capture members of the guerilla forces and their family members. Crispulo "Ising" Atillo is captured and narrowly escapes execution.

Crispulo "Ising" Atillo discovers the body of Teodoro "Doring" Saavedra in the Buhisan River as he escapes Cebu City to reunite with Vicente "Inting" Atillo and Delfin Lopez outside Cebu City.

February 12, 1944 General Douglas MacArthur officially recognizes the guerilla forces throughout the Philippine Islands and begins providing operational support. Captain James Cushing is promoted to Lieutenant Colonel by General Douglas MacArthur retroactive to January 22, 1944, and appointed sole commander of the CAC, 8th Military District, U.S. Army Forces in the Far East (USAFFE).

February 1944 Vicente "Inting" Atillo and Delfin Lopez are transferred from the Combat Company of the 85th Infantry Regiment under Lt. Rogaciano "Popoy" C. Espiritu to "A" Company, 1st Battalion of the 87th Infantry Regiment under Lt. Col. Abel Trazo.

U.S. Navy submarines begin supplying the Cebuano guerillas with weapons, ammunition, medicine, and long-range radio equipment.

March 31, 1944 Admiral Mineichi Koga, the Commander in Chief of the Imperial Japanese Navy's Combined Fleet, and Rear Admiral Shigeru Fukudome, Koga's Chief of Staff, depart Korkor, Palau to Davao City on the Island of Mindanao, Philippines. Admiral Fukudome boards the plane with the secret military plan, Combined Fleet Secret Operations Order No. 73, the "Z Plan."

Shortly after takeoff, the planes encounter a tropical storm. Admiral Koga's seaplane crashes into the Philippine Sea east of Mindanao, killing everyone on board.

April 1, 1944

The second plane, carrying Admiral Fukudome, attempts to change course but crashes in the Bohol Strait off the coast of Cebu near Barrio Balud.

Ricardo Bolo, a Lieutenant of the Volunteer Guards (VG) in Barrio Balud, his younger brother Edilberto, along with their neighbor Valeriano Paradero, discover the survivors of the crash as they neared the beach. Lieutenant Bolo turns the prisoners over to Teopisto Tangub, the VG Commander of Barrio Sangat, who transports the prisoners to Lt. Col. Cushing at the CAC headquarters at Tabunan.

Pedro Gantuangko and his neighbor Rufo "Opoy" Wamer, retrieved the Z-Plan from the wreckage.

April 2, 1944

Col. Seiichi Ohnishi, Commander of the Ohnishi Butai, initiates a massive drive into the interior of Cebu to locate and recover the Japanese prisoners.

April 3, 1944

Teopisto Tangub arrives with the Japanese prisoners at the Command Post of "A" Company, 1st Battalion of the 87th Infantry Regiment, the unit of Vicente "Inting" Atillo and Delfin Lopez. Tangub departs for Tabunan.

Gantuangko turns the Z-Plan over to Corporal Norberto "Berting" Varga, a local member of the

Cebuano guerillas, who begins the trek to deliver the documents to Lt. Col. Cushing.

April 8, 1944 Tangub arrives at the CAC headquarters at Tabunan and turns the Japanese prisoners over to Lt. Col. James Cushing. Corporal Norberto "Berting" Varga arrives with the Z-Plan.

April 9, 1944 The Ohnishi Butai assault the CAC headquarters at Tabunan. Cushing and the Cebuano guerillas evacuate and escape to Kamungayan, where they established a temporary camp.

April 10, 1944 Cushing negotiates the release of the Japanese prisoners to Lt. Col. Ohnishi in exchange for the lives of innocent Filipino civilians being killed by Japanese troops.

April 15, 1944 Cushing sends the Z-plan to Col. Andrews in Southern Negros. They arrive on April 28, 1944.

April 28, 1944 The Z-plan is turned over to Col. Andrews in Southern Negros, Philippines.

May 11, 1944 The Z-Plan is turned over to Lt. Cmdr. Francis David Walker, Jr, Commanding Officer of the USS Crevalle, a U.S. Navy submarine.

May 21, 1944 The USS Crevalle arrives in Australia and the Z-Plan is transported to General MacArthur. Allied commanders exploit and use the secret Japanese military plans to develop a counter-offensive strategy that ultimately leads to the liberation of the Philippine Islands and the defeat of Japanese forces in the Pacific.

September 3, 1944	Governor Hilario "Dodong" Abellana is captured by the Japanese Kempeitai.
October 17, 1944	Allied forces led by General Douglas MacArthur executed Operation King Two and began conducted amphibious assaults on the island of the Leyte Gulf.
October 20, 1944	Allied forces land on Leyte Island. General MacArthur and Philippine President Sergio Osmeña, come ashore at Palo Beach, Leyte, Philippines, fulfilling MacArthur's promise to return to the Philippines and liberate the islands from the Empire of Japan.
December 26, 1944	Allied forces complete the first phase of the operation and recaptured the island of Leyte.

Allied Forces, Leyte Island, Philippines, 1944.

January 15, 1945	Governor Hilario "Dodong" Abellana is executed by the Japanese Kempeitai in Cebu City, Philippines.
March 26, 1945	Major General William H. Harold launches Operation Victor II. Allied forces consisting of the Americal Division's 132nd and 182nd Infantry

                        Regiment land at Talisay Beach, Cebu, Philippines.

March 27, 1945   Allied forces advance into Cebu City, Philippines.

                        Vicente "Inting" Atillo is reported killed in action (KIA) by Lt. Col. Cushing. Atillo miraculously survives and continues fighting.

Allied Forces assault, Talisay Beach, Cebu, Philippines.

March 28, 1945   Allied forces and the Cebuano guerillas secure the port of Cebu and capture Lahug Airfield on Mactan Island, Philippines.

March 29, 1945   Allied forces and the Cebuano guerillas capture and liberate the barangay of Pari-an, Cebu, Philippines.

April 7, 1945   Allied forces and the Cebuano guerillas capture and liberate the barangay of T. Padilla, Cebu, Philippines.

April 8, 1945   Allied forces and the Cebuano guerillas secure the remaining areas of Cebu City, Philippines.

April 13, 1945   Allied forces and the Cebuano guerillas conduct a two-pronged division level attack against the

Japanese and force the remaining Japanese troops into the northern mountains of the island.

June 20, 1945 The Americal Division ceases operations and withdraws to Cebu City, Philippines.

July 2, 1945 The Cebuano guerillas capture the final Japanese stronghold, Camp 8, in the barangay of Minglanilla.

July 26, 1945 Allied Forces issue the Potsdam Declaration calling for the immediate, unconditional surrender of Japan.

August 6, 1945 The United States drops an atomic bomb on the Japanese city of Hiroshima.

American forces personnel examine a "bolo" knife belonging to an Igorot, August 7, 1945.

August 9, 1945 The United States drops a second atomic bomb on the Japanese city of Nagasaki.

Nagasaki, Japan, 1945.

August 15, 1945   The Empire of Japan announced its surrender, bringing World War II to an end.

Vicente "Inting" Atillo and Delfin Lopez are honorably discharged from "A" Company, 1st Battalion of the 87th Infantry of the United States Armed Forces Far East (USAFFE).

August 19, 1945   Lt. Gen. Kataoka Tadasu, Commanding General of the 1st Division of the Japanese Imperial Army, arrives in Cebu City, Philippines, to begin surrender negotiations.

August 28, 1945   Lt. Gen. Kataoka Tadasu surrenders the Island of Cebu and all Japanese forces to Maj. General William H. Arnold of the 23rd Infantry Division in an open field near Ilihan, Cebu, Philippines.

September 2, 1945   The Empire of Japan formally surrenders to General Douglas MacArthur on board the USS Missouri in Tokyo Bay, officially ending World War II.

Surrender of the Empire of Japan, 1945.

November 1945 Lorenzo "Tatay Ensong" Saavedra dies peacefully at his home in Mambaling of natural causes at the age of 93.

February 18, 1946 U.S. President Harry S. Truman and the U.S. Congress rescinded the citizenship and veteran's benefits provided by the Second War Powers Act and pass the Rescission Act of 1946 (Public Laws 79–301 and 79–391; 60 Stat. 6 and 60 Stat. 221). The Rescission Act of 1946 did not consider the wartime service of the Filipinos to be active military service and, therefore, did not qualify for benefits.

April 23, 1946 Sergio Osmeña, Sr. founder of the Nationalista Party loses the presidential election of the Commonwealth of the Philippines to Manuel Roxas, co-founder of the Liberal Party, leaving political and economic control of Cebu City to the family of Philippine Senator, Mariano Jesús Cuenco, Sr.

May 28, 1946 Manuel Roxas is inaugurated as the President of the Commonwealth of the Philippines.

President Manuel Acuña Roxas (January 1, 1892 - April 15, 1948)

July 4, 1946 The United States grants the Philippines total independence.

April 15, 1948 Manuel Roxas, the President of the Philippines dies of a heart attack after giving a speech before the United States Thirteenth Air Force.

April 17, 1948 Vice-President Elpídio Rivera Quiríno is sworn in as the President of the Philippines.

November 13, 1951 President of the Philippine Senate, Mariano Jesús Cuenco, Sr. loses his seat to Senator Quintin Paredes, collapsing the political and economic bond between the Cuenco family and Elpídio Rivera Quiríno, President of the Philippines.

December 30, 1951 Sergio "Serging" Osmeña, Jr. defeats Mariano Cuenco, Jr. and is elected provincial governor of Cebu and Mayor of Cebu City. His election restores political and economic control of Cebu City to the Osmeña family. This leads to greater tension between the Cuenco and Osmeña families

	and the political and economic control of Cebu to include a monopoly on the Port of Cebu.
December 1951	Venancio "Anciong" Bacon decides to disassociate himself from the Doce Pares Club and establish a new club focused on the Saavedra style and the combat applications of Filipino eskrima.
May 1952	Venancio "Anciong" Bacon and Delfin Lopez fight in Tina'an, Cebu, Philippines. The fight is caused by a dispute over the use of the Doce Pares Club name and Anciong's decision to elect himself the head instructor of his proposed club. Lopez concedes to Anciong and surrenders.
June 28, 1952	Venancio "Anciong" Bacon organized a gathering of students in Mambaling and formally announces he is leaving the Doce Pares Club. Anciong founds the Balintawak Self Defense Club along with Vicente "Inting" Atillo and Delfin Lopez. The decision to leave the Doce Pares Club and establish a rival club prompts an intense and legendary rivalry that lasts for decades.
April 9, 1953	President of the Philippines, Elpidio Quirino replaces Dr. Jose Rodriguez and appoints Vicente del Rosario as the major of Cebu City. Rosario begins campaigning in Cebu City for the reelection of President Quirino
May 11, 1953	Vicente del Rosario demonstrated his control of Cebu City by summarily dismissing forty-three detectives of the Cebu City Police Secret Service Force, including Delfin Lopez. Their dismissal is arguably an attempt to manipulate the ranks within the Cebu City Police and remove detectives who

posed a threat against the newly appointed Mayor and the re-election of President Quirino.

1953   Vicente "Inting" Carin of the Doce Pares Club shoots Delfin Lopez at a reelection campaign event for President Quirino in the barangay of Tisa in Cebu City, Philippines. Lopez narrowly survives. Carin is arrested and charged for the shooting and attempted murder of Delfin Lopez and incarcerated in Bilibid Prison in Muntinlupa, Philippines. The incident greatly intensifies the rivalry between the Doce Pares Club and Balintawak Self Defense Club.

Delfin Lopez founds his own security company, the Delfin N. Lopez Security Agency.

November 10, 1953   President of the Philippines, Elpídio Quiríno loses reelection to former Philippine Defense Secretary, Ramon Magsaysay.

December 6, 1953   President Magsaysay replaces Vicente Del Rosario with Dr. Jose Chiong Veloso Rodriquez as the Cebu City mayor.

April 17, 1954   Atty. Democrito T. Mendoza co-founds the Associated Labor Union (ALU) in Cebu City, Philippines. Mendoza is a friend and student of Venancio "Anciong" Bacon and a former Balintawak Self Defense Club classmate of Delfin Lopez.

October 1956   The ALU launches its first strike against the Visayan-Cebu Terminal Company paralyzing the port of Cebu. The Cuenco family hires Delfin Lopez to provide crucial strike-busting muscle and protect strikebreakers for the company so cargo operations can continue throughout the ALU strike.

October 6, 1956 Delfin Lopez is dubiously arrested by the Cebu City Police and jailed for a series of alleged murders dating back to 1947. His home is raided by the Cebu City Police under direct orders from Sergio Osmeña, Jr. Osmeña then facilitated temporary control over dock operations in the Port of Cebu and recognized Mendoza and the ALU as the exclusive bargaining agent for all workers in the port. The arrest is made political "bail" by the opposing political factions, and Lopez is eventually released, and all criminal charges are summarily dismissed.

1956 Delfin Lopez fights Florencio "Inciong" Lasola at the residence of Otillo "Lolo" Larawan, in the Lagtang barangay of Talisay City, Cebu, Philippines. Lasola is easily defeated by Lopez.

August 26, 1963 James Cushing passes away of a heart attack at 53 years old while on an inter-island transport. He is interred in the Heroes' Cemetery in Manila, Philippines.

May 1964 Crispulo "Ising" Atillo fights and defeats Lauriano "Lauren" Sanchez in the barangay of Tisa, Cebu City, Philippines.

August 1964 Crispulo "Ising" Atillo fights and loses to Antonio Irogirog in Cebu City, Philippines.

September 24, 1964 Delfin Lopez is killed by Emiliano Berania while attempting to calm an ALU labor strike in Cebu City, Philippines. Venancio "Anciong" Bacon is suspected of collaborating with the ALU and providing information that lead to Lopez's death.

Berania is found not guilty and acquitted of the murder despite there being several witnesses. Berania is mysteriously killed in Mindanao, Philippines.

1965 Venancio "Anciong" Bacon is arrested for the murder of Pio Deparine and incarcerated in the criminal detention facility at Camp Crame, headquarters of the Philippine Constabulary of the Armed Forces of the Philippines (AFP) in Quezon City, Philippines.

The Balintawak Self Defense Club begins to dissolve into several separate clubs and training groups.

September 23, 1972 President Ferdinand Marcos declares martial law throughout the Philippines and orders the arrest of Senator Benigno "Ninoy" Aquino, Jr. on charges of murder and subversion.

President Ferdinand Marcos declares martial law.

January 1975	General Fabian C. Ver, the Chief of Staff of the Armed Forces of the Philippines, along with Romeo C. Mascardo, found the National Arnis Association of the Philippines (NARAPHIL).
1975	Venancio "Anciong" Bacon is paroled from the Criminal Detention Facility at Camp Crame and relocates to Iligan City in Northern Mindanao, Philippines. He regularly returns to Cebu City to satisfy the conditions of his parole and visit his students.
	Venancio "Anciong" Bacon and Crispulo "Ising" Atillo participate in a sparring match that turns into an actual fight in Mambaling behind the Lady of Lourdes Parrish Church in Punta Princesa, Cebu, Philippines.
1975	Dionisio "Diony" Cañete founds the Cebu Eskrima Association (CEA) in Cebu City, Philippines.
April 24, 1975	Vicente "Inting" Atillo and Crispulo "Ising" Atillo found the New Arnis Confederation of the Visayas and Mindanao (NAC). The name is later changed to the Philippine Arnis Confederation (PAC).
	The PAC becomes the first eskrima organization of Cebu, Philippines, to affiliate with NARAPHIL.
November 25, 1977	Senator Benigno "Ninoy" Aquino, Jr. is dubiously convicted of murder and subversion by a military tribunal and sentenced to death.
March 24, 1979	The Cebu Eskrima Association (CEM) in association with NARAPHIL sponsors the First

National Open Arnis Championships in Cebu City, Philippines. Ciriaco "Cacoy" Cañete deceitfully wins the tournament and is declared the National Champion.

Cacoy issues a public challenge to Crispulo "Ising" Atillo and all of Balintawak eskrima reigniting a decades old rivalry.

August 19, 1979 NARAPHIL sponsors the First National Invitational Arnis Tournament at the Philippine National College Gymnasium in Manila, Philippines. Ciriaco "Cacoy" Cañete deceitfully wins the tournament.

1979 Venancio "Anciong" Bacon's health begins to decline due to years of poor health and smoking.

November 1, 1980 Venancio "Anciong" Bacon passes away on at the age of 69 and is buried in the San Nicolas Catholic Cemetery in Cebu City, Philippines.

January 17, 1981 President Ferdinand Marcos lifts martial law in the Philippines.

July 1, 1983 The Freeman Daily newspaper headlines Crispulo "Ising" Atillo challenges Ciriaco "Cacoy" Cañete to a fight without any protective equipment "anytime, anywhere."

July 3, 1983 The Freeman Daily newspaper publishes that Ciriaco "Cacoy" Cañete accepts Crispulo "Ising" Atillo's challenge.

August 21, 1983 Philippine Senator, Benigno "Ninoy" Aquino Jr. is assassinated at the Manila International Airport moments after his plane lands upon returning from exile in the United States.

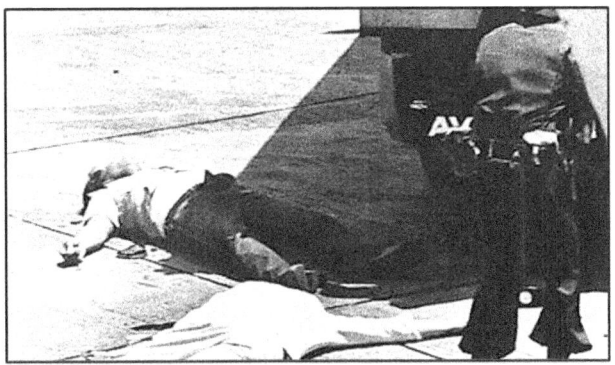

Assassination of Philippine Senator, Benigno "Ninoy" Aquino Jr., August 21, 1983.

September 16, 1983 Crispulo "Ising" Atillo and Ciriaco "Cacoy" Cañete review the contract for the match and sign agreeing to the rules and regulations.

September 17, 1983 Crispulo "Ising" Atillo fights Ciriaco "Cacoy" Cañete in the last official challenge match. The match ends in reported controversy.

September 20, 1983 Crispulo "Ising" Atillo is ordered by the Philippine Constabulary to participate in a fake rematch with Ciriaco "Cacoy" Cañete to divert anti-Marcos protestors from a rally protesting the assassination of Senator Benigno "Ninoy" Aquino.

Crispulo "Ising" Atillo is advised if he does not participate in the ruse he will be killed.

September 21, 1983 Crispulo "Ising" Atillo and Ciriaco "Cacoy" Cañete participate in the ruse held at the Cebu Coliseum, Cebu City, Philippines. Crispulo "Ising" Atillo is declared unfit to fit. The fight is called off.

Anti-Marco protest marches protesting the assassination of Philippine Senator, Benigno "Ninoy" Aquino Jr.

August 7, 1989   Vicente "Inting" Atillo falls ill and is diagnosed at the Cebu City Medical Center with Pneumonitis and Atherosclerosis, after which his health begins to steadily decline.

October 26, 1990   The U.S. Congress passes the Immigration Act of 1990. The updated act allowed for the naturalization of Filipinos who served on active duty as members of the USAFFE, Philippine Army, Philippine Scouts, and recognized Guerilla Units between September 1, 1939, and December 31, 1946.

August 19, 1991   Vicente "Inting" Atillo completes the U.S. naturalization process and is scheduled to participate in his final interview at the U.S. Embassy in Manila. His health declines and he is confined to his bed.

June 8, 1993   Vicente "Inting" Atillo peacefully passes away at home from heart failure.

January 26, 2001	Crispulo "Ising" Atillo travels to the United States to conduct a demonstration at the World Eskrima Kali Arnis Federation (WEKAF) tournament in Northern California and teach a series of seminars on Balintawak eskrima. He resides with Max Pallen in San Leandro, California.
March 2001	Crispulo "Ising" Atillo moves to Buena Park, California to live with Ramon Rubia.
June 2001	Crispulo "Ising" Atillo moves from Ramon's residence in Buena Park, California, to Loma Linda, California, to live with his mom's first cousin, Lucy Abela, Venarendo Ylaya, and Alvin Ylaya.
December 2002	Crispulo "Ising" Atillo relocates to Beaumont, California to live with Dr. Jesse Devera and establishes the World Headquarters of Atillo Balintawak Eskrima - Original Saavedra Style.

Glossary of Terms

abaga	The human shoulder.
abaniko	A fan-like striking movement with a stick. See also *pamaypay*.
abatan	To intercept or hit without warning.
abante	To advance or move forward.
abecedario	A term meaning "ABC Diary." Commonly used to describe fundamental movements and techniques in Filipino eskrima.
abtik	Ready to take action or to move quickly.
abyerta	A term meaning "open" used to describe a position of the stick away from the body.
adlao	Sun.
agak	The act of coaching or guiding another person.
agaw	To take ahold of something or grab. Commonly used in eskrima to describe a disarm technique.
agimat	An amulet or talisman. See also *anting-anting*.
agtang	The brow or human forehead.

agwat	Distance.
akay	To guide or direct.
aldabis	A backstroke or upward backhand strike.
amara	The twirling or swinging of a stick to demonstrate speed, power, and skill. Also used in Filipino eskrima to describe a striking combination.
Ang estilo nimo ay bati.	A phrase meaning "Your style is ugly." Commonly used by GGM Atillo to suggest you need more practice.
antas	A *level* or *grade*. Commonly used in eskrima to denote a level or rank.
antaw	Objects that are visible from a distance. Often used to describe long-distance techniques used in Filipino eskrima.
anting-anting	A trinket or piece of jewelry usually hung around the neck, thought to bring good luck, and give magical protection against evil or disease.
apak	To step or step on.
apapangig	The human jaw.
aping	The cheek of the face
armas	A Spanish term meaning arms or armaments; weapons.

armas de mano	A Spanish term meaning *hand armaments* or weapons held in the hand.
arnés	A Spanish term that literally translates to *harness*. Often used in Spanish sword fighting to refer to protective armor.
arnis	A term commonly used to refer to the Filipino martial arts and national martial art of the Philippines. An adaptation of the Spanish word *armas,* meaning *arms* or *weapons*. Often mistakenly confused with the Spanish term *arnés*. See also *arnis de mano*.
arnis de mano	An adaptation of the Spanish term armas de mano, or "hand weapons." Commonly used to refer to the Filipino martial arts. See also *arnis*.
atras abante	A back and forth movement or footwork that retreats then advances forward.
baba	The human mouth
babag	To fight or battle.
bagtak	The human calf muscle
bahad	To threaten or verbally assault. Commonly used to describe a duel or grudge match between rival eskrimadors. See also *deathmatch* and *juego todo*.
baksing	Cebuano term for the sport of boxing.
balaraw	A dagger with a double-edged blade.

bali   A bone fracture or break.

Balintawak   The name of a small municipality in the area of Caloocan City where it is believed an outcry was held signifying the beginning of the Philippine Revolution against the Spanish by members of the Katipunan led by Andrés Bonifacio in 1896. Also referred to as the *Cry of Pugad Lawin* due to debate over the exact location of the event. The name of a small side street in Cebu City, Philippines, and site of the Balintawak Self Defense Club.

Balintawak Eskrima   A generic term used to describe the many styles and systems of Filipino eskrima originating from the Balintawak Self Defense Club founded in 1952 by Venancio "Anciong" Bacon along with other well-known eskrimadors.

bali-bali   Changing grips or changing hands with the stick.

balisong   A fan-shaped knife originating from the Batangas region of the Philippines.

baltak   A forceful and sudden pull.

balutin   To entangle or wrap up. See also *hubud*.

banda y banda   The side to side motion of the stick

bantay   To guard and protect yourself.

bangkaw	A native spear of the Philippine Islands. Also used to describe long-staff techniques in Filipino eskrima. See also *sibat*.
bantay	To guard or protect.
barangay	The smallest political unit in the Philippines. Similar to a town or village.
baraw	A short knife or dagger. See also *daga*.
bastón	A Spanish term for a stick used in Filipino eskrima. See also *olisi*.
batak	To pull an object or person.
bat-ang	The human hip
batikan	An expert of a skill.
batukan	The back of the human neck.
bayani	A hero.
bibig	Mouth.
bigwas	To strike someone with a first. See also *suntok*.
bihasa	A term used to describe someone experienced or well-trained.

bikas	An attitude or posture.
binali	Being inverted or opposite such as a mirror image of oneself.
binti	The human leg.
Bisaya	A Visayan or native of the Visayan region of the Philippine Islands.
bitin	A snake or serpent.
bitiis	A term that generally refers to the entire leg, however it is commonly used to describe the portion of the leg between the knee and the ankle.
boksing	A Cebuano term meaning to fight with the fists or the sport of boxing. See also *suntukan*.
bolo	A machete or long knife commonly used throughout the Philippines
braso	The human bicep
brazos cruzados	A Spanish term that refers to crossed arms.
budō	A Japanese term describing the Japanese martial arts. Literally translates to mean "Martial Way" and is thought of as the "Way of War."
bubod	To coil or entangle. See also *higot*, *hubud*, and *lubud*.

bugnó　　　　　　To fight or do battle.

buhi　　　　　　　To survive or be alive.

buhok　　　　　　Human hair.

bukog　　　　　　A human bone.

bukton　　　　　　The human arm or forearm.

bunal　　　　　　A stick used to strike.

bunŏ　　　　　　 Also commonly used term used to describe wrestling techniques.

bũno　　　　　　 To assassinate or murder.

buntal　　　　　　To box or strike someone with the first.

butangan　　　　　To hit or beat someone mercilessly.

butáng kabilin　　An heirloom handed down by tradition.

cabra　　　　　　A ripping thrust with the point of the stick.

cadena　　　　　　A chain. See also *kadena*.

Cárcel del Distrito　　A Spanish term used to describe a district jail.

Cebu	An island in the Visayan region of the Philippine Islands. Commonly believed to be the heart of the Filipino art of eskrima.
Cebuano	A native of the island of Cebu in the Visayan region of the Philippine Islands. Also, the native language of the islanders.
Cebuano Eskrima	A term used to generally describe the styles and systems of eskrima originating from the Philippine island of Cebu and the Cebuano people.
check	A term used in eskrima to describe the use of the left hand to control and monitor the opponent's hand. Concept similar to checking in chess. See also *tapi-tapi*.
combaté	The act of fighting or engaging in combat with another person or military force.
Combat Judo	A common term used throughout the Philippine Islands to refer to the empty hand self-defense techniques of the Filipino martial arts. Not directly related to the Japanese art of judo.
compadres	A way of addressing or referring to a friend or companion. Commonly used by Filipino veterans of World War II to refer to their fellow compatriots. See also *paré*.
corridas	A phase of training in Balintawak eskrima using random strikes and counters. See also *korridas*.
corrido	A metrical story about the life and adventures of a person usually sung in the accompaniment of a guitar. See also *korido*.

corto	A short distance or close range. See also *korto*.
corto kurbada	Close-range curving strikes used in the Doce Pares styles of Filipino eskrima.
corto linear	A common term used to describe a close-quarters method of eskrima with short lines of attack and defense.
crossada	A term often used to describe a cross-arm style or position in Filipino eskrima.
crusada	A crusade. See also *krusada*
cruzados	Crossed. See also *brazos cruzados*.
cuentada	The action of computing or calculating. In Balintawak eskrima, it is the process of calculating and counter-attacking an opponent's moves like a game of chess. See also *kuentada* and *kuwentada*.
cuerda	A rope or string.
daga	A knife or dagger. See also *baraw*.
dagi	To bump or push with the elbow or shoulder.
dalunggan	The human ear,
dakma	To grab or seize with the hand.
damba	To jump and kick.

dangal	Honor.

dapit	A person's position in relation to another person or an object.

datu	A tribal leader or Chieftain, particularly in the Southern Philippines.

dawaton	To receive.

daya	A term that refers to someone who cheats or defrauds another through lies and deceit.

deathmatch	A duel or grudge match between rival eskrimadors where no safety equipment is worn. See also *bahad* and *juego todo*.

de cadena	A chain or series of things or actions linked together. See also *de kadena*.

de kadena	A chain or series of things or actions linked together. See also *de cadena*.

de cuerdas	A string or series of things or actions.

dikit	To collide violently or come in immediate contact with another object.

dila	The human tongue.

diwa	Spirit.

doble	The act of doubling or copying something.

doble bastón	A term used to describe the use of two sticks in Filipino eskrima. See also *doble olisi*.
doble olisi	A term used to describe the use of two sticks in Filipino eskrima. See also *doble baston*.
Doce Pares de Francia	Spanish translation of the *Twelve Peers of France*, the Paladins or twelve foremost knights of King Charlemagne's court. Often mistakenly translated as the *Twelve Pairs of France*.
dughan	The human chest.
dugo	Human blood.
dumala	The act of being in charge or in control.
dumog	The act of fighting or to tussle. Also commonly used to describe native Filipino wrestling. See also *layog*.
dunggab	To stab.
dunggal	To jab or poke.
dunggan	The human *ear*
eksamin	A test or examination. See also ***pagsubok***.
engganyo	To lure or entice someone into doing something through a false promise or action. A faking motion in the art of eskrima.
escrime	A French term for European sword fighting and the sport of fencing.

escrimeur	A French term for a practitioner of the art of fencing or Filipino eskrima. See also *eskrimador*.
esgrima	A Spanish and Portuguese term for European sword fighting and the sport of fencing.
Eskrido	A style of Filipino martial arts founded by Ciriaco "Cacoy" Cañete in 1951.
eskrima	A Filipino term for European sword fighting and the Filipino art of stick-fighting. See also *estukada*.
eskrimador	A practitioner of the Filipino art of eskrima.
espada	Sword.
espada y daga	Sword and dagger.
estilo	A style or way of expressing something characteristic of a particular person or group of people.
estocada	To thrust. See also *pagduso*.
estudiante	A student.
estukada	A filipino term used to describe Spanish fencing and the Filipino art of stick-fighting. See also *eskrima*.
estilo bitin	Snake-style of eskrima.
estilo korto	A close-quarter or close-style of eskrima. See also *corto mano*.

estilo kadena A chain-style of eskrima or a series of techniques linked together.

estilo layo A long-range style of eskrima. See also *largo mano*.

falla A fine paid to exempt oneself from forced labor during the Spanish colonial era of the Philippine Islands. See also *la multa*.

fencing European sword-fighting and the sport of fighting with swords.

gabay To guide or lead.

galang To show courtesy or respect.

garrote A handheld length of chain, rope, or wire used to strangle a person. A device used by Spanish authorities to execute prisoners sentenced to capital punishment.

gawas A physical position on the outside.

goon A common term used to describe a person who is a bully or thug, especially one hired to intimidate others.

Grandmaster A rank or title given to a person who is the head instructor or highest-ranking authority of a given martial arts style or organization.

Guardia Civil A branch of the Spanish Army organized under the Spanish colonial government in the Philippines responsible for enforcing local laws.

gunting Scissors or a cross-cutting motion.

guro	An academic teacher. A term often used to describe or denote rank as an instructor in the Filipino Martial Arts. See also *magtutudlo*.
hagibis	The sound produced when an object such as a stick quickly passes by.
haginit	The cracking sound produced when a stick strikes another object.
hahabol	To run after or chase someone.
hakbang	To step or walk.
halabas	A long bolo or the act of striking and cutting with the bolo.
halibas	To strike or lash with a stick.
haltak	To jerk or forcefully pull.
hambalos	To strike right and left with a stick or other object.
hamon	To challenge someone.
hampas	To hit or strike, particularly as a form of punishment.
hapak	To hit or strike.
harang	To block or intercept a blow.
hawak	The hips or waist.

hawan	To clear or remove an object.
harakiri	A form of Japanese ritual suicide by disembowelment. See also *seppuku*.
hasa	To process of practicing to improve or to be well-trained or experienced.
hatak	To pull with the hand.
hataw	To hit with the hand or by an object held in hand.
higot	To tether or tie something up. See also *bubod*, *hubud*, and *lubud*.
hiklat	To forcibly widen or open by creating space.
hilagpos	To loosen or untie.
hubad	To unravel or untie. See also *higot*, *bubod*, and *lubud*.
hulog	To trip or fall down.
igo	Sufficient or good enough.
ikis	A crossing or X-pattern motion.
ilag	To evade or parry.
ilok	Armpit.

ilong	Nose.
inat	To stretch the body.
insurrection	A violent uprising against an authority or government. See also *pagsukol*.
iring ug iro	A Cebuano term meaning "cats and dogs." An offhanded term used to describe contentious training at the Labangon Fencing Club.
isog	To be brave or courageous.
iwa	To slash or cut with a blade.
iwas	To move out of the way or avoid.
judo	A Japanese martial art and sport created by Jigoro Kano in 1882 that emphasizes physical education and personal development.
jugo todo	A Spanish term that means to *play everything* or *game all*. It is commonly used in Filipino eskrima to refer to a deathmatch where no safety equipment is worn and anything goes. See also *bahad* and *deathmatch*.
jujutsu	Japanese martial art and a method of close combat on the battlefield.
juramentado	An early 20th century native of the Philippines who swears allegiance to Islam and pledges a willingness to die for the Islamic faith while engaged in combat.
kaaway	An enemy or adversary.

kabilin	Practices that are handed down by inheritance or tradition.
kadena	A chain. See also *cadena*.
kahig	A scratch.
kalaban	An opponent.
kalang	A wedge.
kalasag	A shield.
kalayaan	Freedom.
kali	A term used to describe the Filipino Martial Arts. In particular the early indigenous arts from the Central and Southern Philippines.
kalinangan	Tradition and culture.
kalis	A long-pointed sword.
kalit	Sudden or abrupt.
kaluban	A sheath or scabbard.
kalmutin	To scratch or claw.
kamay	A Tagalog term for the human hand.

kamot	A Cebuano term for the human hand.
kampay	To swing or stroke.
kapit	To grip or hold firmly with the hand.
kapit sa kamot	To hold the hand firmly.
karate	A Japanese martial art developed in the Ryukyu Kingdom, primarily the island of Okinawa, from the indigenous Ryukyuan martial arts.
kata	A Japanese term used to describe the choreographed pattern of movements practiced in Japanese martial arts.
katalo	An antagonist, adversary, or opponent.
Katipunan	A Philippine revolutionary society founded by anti-Spanish colonialism Filipinos in 1892. Also known as the *Katipunan ng mga Anak ng Bayan* (Supreme and Venerable Society of the Children of the Nation) or KKK.
Katipuneros	Members of the Katipunan.
katungaali	An antagonist or rival.
Kempeitai	A military police branch of the Imperial Japanese Army that served as both a conventional and secret military police force.
kerida	A term for a mistress.

kilay	The eyebrow or brow line of the human forehead.
kiling	To lean or slant the body.
kinaadman	Wisdom and experience.
kolehiyo	A college or college-level education.
kontra	An enemy.
korido	A metrical story about the life and adventures of a person usually sung in the accompaniment of a guitar. See also *corrido*.
korridas	A phase of training in Balintawak eskrima using random strikes and counters. See also *corridas*.
korto	A short distance or close range. See also *corto*.
kris	A wavy or straight double-edged blade generally believed to originate from Indonesia.
krusada	A crusade.
kuko	Fingernail.
kumagko	Thumb.
kumingking	The little finger of the human hand.
kumot	The fist.

kunsi	A bolt or latch. Commonly used to describe a joint lock or locking position in eskrima.
kusog	Strength.
kutsilyo	A generic term for an edged weapon or tool.
kuwenta	To compute something. Also spelled *kwenta*.
kwentada	The action of computing or calculating. In Balintawak eskrima, it is the process of calculating and counter-attacking an opponent's moves like the game of chess. Also spelled *kuwentada* and *cuentada*.
Kwentada Eskrima	The Balintawak eskrima style of Venancio "Anciong" Bacon.
labnot	To snatch suddenly with the hand.
labo-labo	Chaos and turmoil often associated with a brawl, uprising, or riot.
lagay	To slide.
lag-lag	To fail or be unsuccessful.
lakan	A tribal leader or person of high rank.
lakang	A step when walking or running.
lakang sa atubang	Step to the front or face-to-face.

lakang sa likod	Step backward.
lakang sa likuron	Step to the rear.
lakang sa tuo	Step to the right.
lakang sa wala	Step to the left.
lakaw	To walk.
lalang	To dodge or avoid by deceit.
la multa	A fine or penalty. See also *falla*.
lansis	A trick.
lantay	A wooden bench typically made of bamboo.
larga	A feminine Spanish noun meaning long or lengthy.
largo	A masculine Spanish noun meaning long or lengthy.
largo mano	A common term used to describe the long hand or long-range style or techniques of Filipino eskrima in masculine form. Conversely, the action performed by a female practitioner would be *larga mano*.
latik	A strike that flicks or retracts after impact. See also *witik*.
lawas	The human body.

layo	A Cebuano term for the distance between two objects or people.
layong distansya	A Cebuano term for a long distance between two objects or people.
layog	The act of fighting or to tussle. Also commonly used to describe native filipino wrestling. See also *dumog*.
lechón	A Spanish word referring to a roasted pig.
lihok	Physical action or an object in motion.
lihok sa tuo	Move to the right.
lihok sa taas	Move above.
lihok sa ubos	Move underneath.
lihok sa wala	Move to the left.
likha	To create.
liksi	Mental quickness and agility.
linear	A line between two objects. See also Cebuano *linya*.
linlang	Trickery and deceit.
linya	A line between two objects. See also *linear*.

liog	Neck.
lubud	To unravel or untie. See also *higot*, *hubad*, and *bubod*.
lubot	Buttocks.
luklukan	A throne or seat of honor.
lukso	To jump or leap. See also *lundag*.
lundag	To jump or leap. See also *lukso*.
maestro	A teacher or master. See also *magtutudlo* and *guro*.
magiting	To be brave or courageous.
magtutudlo	An instructor or coach. See also *maestro* and *guro*.
mahabang kamay	A Tagalog term meaning long-hand or long-range techniques in Filipino eskrima.
maharlika	Nobility and upper class.
makabayan	A patriot or someone patriotic.
malaya	Free and independent.
Mambaling	A barangay in Cebu City, Philippines, and home of GGM Crispulo "Ising" Atillo.

Mambaling Style	A term used to describe the style of Filipino eskrima originating from the Saavedra family in the Mambaling barangay of Cebu City, Philippines.
manimbang	Balance.
mano	Spanish term for the human hand.
marangal	A person who is honorable and noble.
matuk-an	To choke. See also, *sumakal and tuk-on*.
medio	The middle or midway point. See also *medyo*.
medyo	The middle or midway point. See also *medio*.
mestizo	A Spanish term used to describe people of the Philippine Islands of mixed race or foreign ancestry.
Modern Arnis	A style of Filipino martial arts founded by Remy Presas as a system of self-defense as well as injury-free method to preserve the older styles of Filipino eskrima.
Moro	A Filipino of the Islamic faith. A *Muslim*.
moro-moro	A type of theatrical stage-play popular in the Philippines during the Spanish colonial period that portrayed the battles between the Moro and Christian inhabitants of the Philippine Islands.
olisi	A walking stick or stick used in the art of eskrima.

orasyón — A prayer.

pagduso — To thrust.

pagsubok — A test or examination. See also *eksamin*.

pagsukol — A violent uprising against an authority or government. See also *insurrection*.

pagsuntok — A Tagalog term for punching with the fists.

pagtatanggol — Defense.

pagtulog — To push someone.

pa-in — To bait or lure.

pakundangan — To show respect.

palad — Palm.

palusot — To excuse. In eskrima, it commonly means to pass a strike unopposed.

paligsahan — A contest or competition.

paltik — A flip or flick of the stick.

pamana — Heritage.

pamaypay	Tagalog term for a *fan*.
pananjakman	A dialectal Filipino term that describes kicking with the foot.
panantukan	A dialectal Filipino term that describes the Filipino adaptation of the sport of boxing.
panata	A vow or promise.
panday	A blacksmith.
pangamba	Fear.
pangamot	A term used to describe the empty hand fighting techniques of the Filipino Martial Arts.
paré	An affectionate Filipino slang term used to describe a close friend. See also *compadres*.
patas	Equal or a tie.
patnubay	To guide.
payapa	Peaceful or tranquil.
payong	Umbrella.
pinuno	Leader.
pinuti	A garrote or long blade.

pisaw	A type of long knife.
pisil	To squeeze.
pitik	To flick with the fingers.
polo y servico	A system of forced labor introduced to the Philippine Islands by the Spanish.
pugay	A tribute, salute, or honor.
pukol	To throw an object.
pulso	The wrist.
puluhan	The shaft or handle of a sword or knife.
punal	A short-bladed weapon.
punta	The point of a knife, sword, stick, spear, etc.
punta y daga	A common term used to describe the *stick and dagger* method of eskrima.
redondo	Round or circular.
retirada	To retreat. Commonly used to describe a retreating style of eskrima.
sabayan	A Cebuano term that refers to two actions occurring at the same time. A simultaneous attack and defense.

sabid	To tangle or entangle.
sabong	Cockfighting. An extremely popular gaming sport throughout the Philippine Islands.
sagang	To shield against or block.
sagi	The sideswipe or bump with the shoulder.
sagisag	A symbol or emblem.
saksak	To thrust or stab. Also commonly known as a grip, whereas the point of the knife is facing upward for thrusting.
Sala de lo Criminal	Spanish criminal court.
salakay	To assault or attack.
sampa	To slap. See also *sampalin*.
sampalin	To slap. See also *sampa*.
sangga	To parry or block.
San Miguel Eskrima	A style of Filipino eskrima founded by Filemon "Momoy" Cañete of the Doce Pares Club.
sapok	To punch. See also *suntok*.
scherma	An Italian term for European sword fighting and the sport of fencing.

seppuku	A form of Japanese ritual suicide by disembowelment. See also *harakiri*.
sibat	A native spear of the Philippine Islands. Also used to describe long-staff techniques in Filipino eskrima. See also *bangkaw*.
sikad	To kick with the foot. See also *sipa*.
sikaran	A Filipino kicking art originating from a children's game in the town of Baras in the province of Rizal.
siko	Elbow.
siniwali	Originating from the word *sawali* for woven bamboo strips, it is commonly used to describe the interweaving patterns and use of two sticks in the art of eskrima.
sipa	To kick with the foot. See also *sikad*.
solo olisi	Single stick.
songab	A finger jab or thrust to the eyes. See also *sunggab*.
subversion	The undermining of the power and authority of an established system or institution.
sulod	To move inside.
sumakal	To choke or strangle.
sumbag	The hit with the hand or fist.

sumbrada	A counter-for-counter style.
sundang	A knife or dagger. A general term referring to bladed weapons and agricultural tools with a bladed edge.
sunggab	To grip or grasp. Also commonly known as a finger jab or thrust.
sungkit	A long pole.
suntok	To hit with the fist.
suntukan	A Tagalog term to fight with the fists or the sport of boxing. See also *boksing*.
suwag	Headbutt.
taal	Native-born.
taas nga kamot	A Cebuano term meaning long-hand or long-range techniques in Filipino eskrima.
tabas	To strike or attack from an angle.
tabon	To cover against.
tadyak	To kick. See also *sipa*.
tadykan	To kick repeatedly or foot fighting.
Tagalog	The national language of the Philippines.

tagpas	To chop down or slash.
taming	A shield.
Tatay	An affectionate term for a father or person who is seen as a father-figure.
tapi-tapi	Meaning "tap-tap" commonly referring to the actions of the checking or non-stick hand in Filipino eskrima.
tigbas	To hack with a blade or stick.
timbang	Weight distribution and balance.
tindi	Intensity.
tiwala	Trust.
trangka	To trap or lock up.
tuhod	Knee.
tuhon	A master.
tuk-on	To smother.
tulak	To shove or push.
tulisan	A bandit.

uway   Rattan.

uyog   To shake.

witik   A flicking or flipping motion of the stick.

yantok   Rattan or cane.

yapak   To step on.

Memories

The following pages include a random selection of photographs of Great Grandmaster Crispulo "Ising" Atillo through the years.

Balintawak Escrima Grandmaster Crispulo "Ising" Atillo — Western Fencing Master & Edwin A. Richar[…]

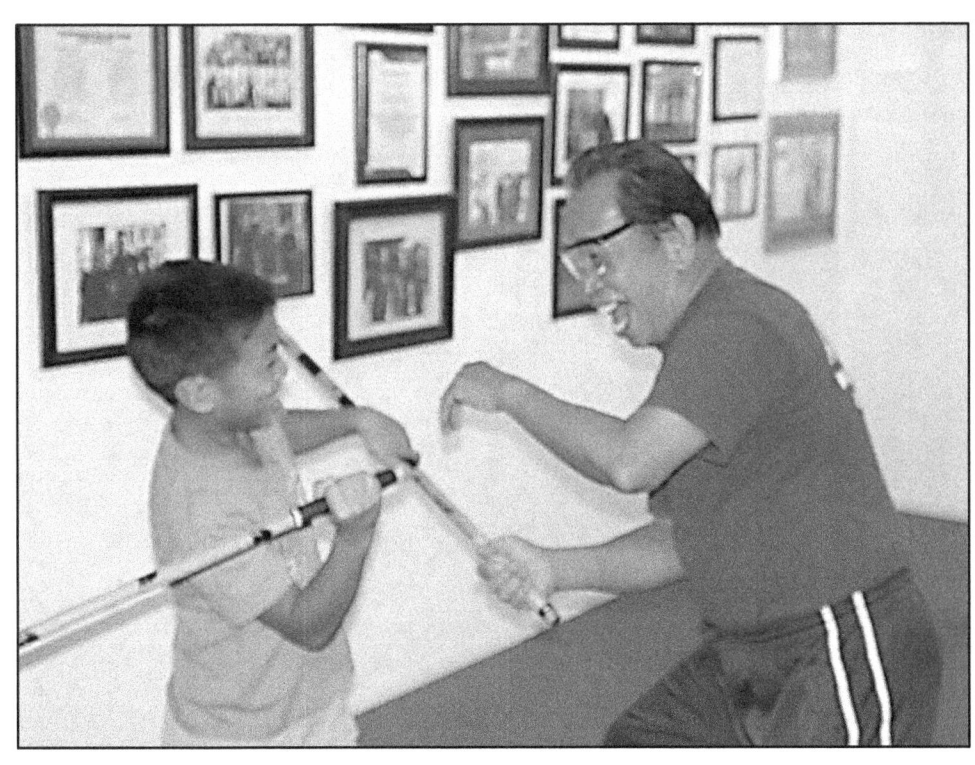

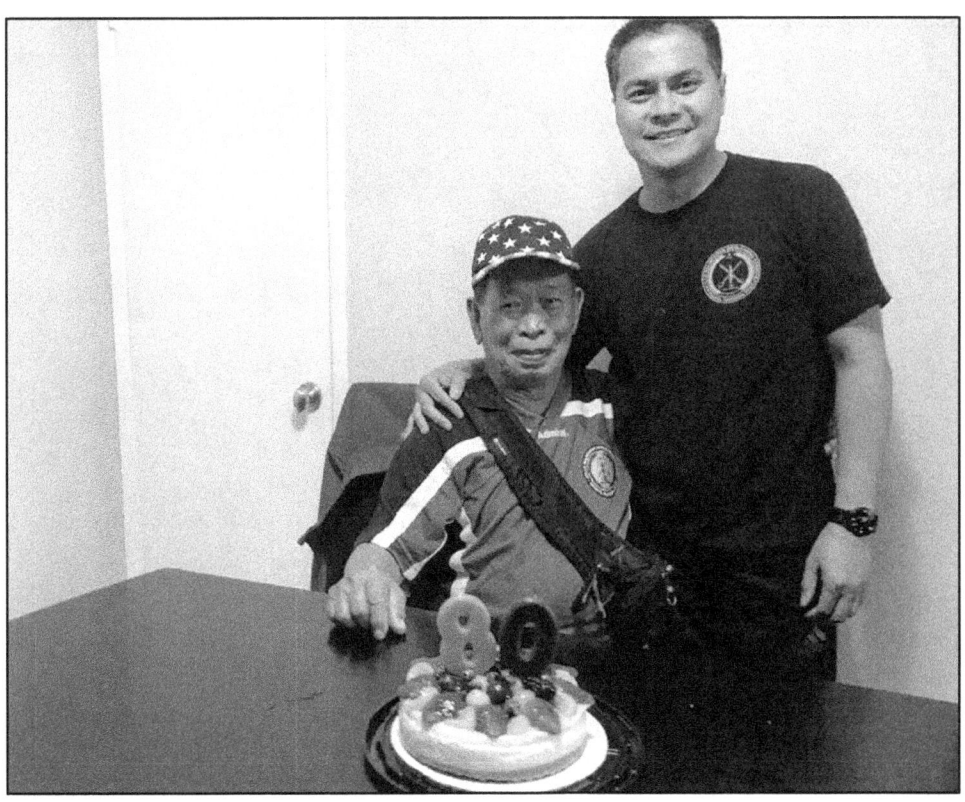

Grand Master Atillo with Dennis & Derrick Dalan

300

Bibliography and References

Abellana, Jovito, *My Moments of War to Remember By*. Cebu City, Philippines: University of San Carlos Press, 2011.

Barreveld, Dirk Jan, *Cushing's Coup*. Havertown, Pennsylvania: Casemate Publishers, 2015.

Borromeo-Buehler, Soledad, *The Cry of Balintawak*. Manila, Philippines: Ateneo De Manila Univ Press, 1999.

Breuer, William B., *MacArthur's Undercover War: Spies, Saboteurs, Guerillas and Secret Missions*. Edison, New Jersey: Castle Books, 1995.

Buot, Sam L. Sr. *Balintawak Eskrima*. Phoenix, Arizona: Tambuli Media, 2015.

Cañete, Ciriaco, *The Challenge Fights of Grandmaster Ciriaco "Cacoy" Cañete*. Los Osos, California: Freehand Publishing, 2014.

Cañete, Dionisio A. Eskrima Kali Arnis. Cebu City, Philippines: Doce Pares Publishing House, Inc., 1993.

Castle, Egerton, *Schools and Masters of Fencing: From the Middle Ages to the Eighteenth Century*. Dover Publications, 2003.

Diego, Antonia and Ricketts, Christopher, *The Secrets of Kalis Ilustrisimo*. Boston, Massachusetts: Tuttle Publishing, 1999.

Draeger, Donn F. *The Weapons and Fighting Arts of Indonesia*. Bunkyo-ku, Tokyo: Charles E. Tuttle Company, Inc., 1972.

Fraguas, Jose M. *Escrima Masers*. Los Angeles, CA.: Empire Books, 2018.

Galang, Reynaldo S. *Warrior Arts of the Philippines*. Roseland, New Jersey: Arjee Enterprises, Inc., 2005.

Giron, Leo M. *Giron Escrima: Memories of a Bladed Warrior*. Los Angeles, CA: Empire Books, 2006

Holmes, Kent, *Wendell Fertig and His Guerrilla Forces in the Philippines: Fighting the Japanese Occupation, 1942-1945*. Mclean, Virginia: McFarland Publishing, 2015.

Ishida, Jintaro, *The Remains of War, Apology and Forgiveness, Testimonies of the Japanese Imperial Army and Its Filipino Victims*. Gullford, Connecticut: Megabooks Company, 2001.

Macachor, Celestino C. and Nepangue, Ned R. M.D., *Cebuano Eskrima: Beyond the Myth*, Manila, Philippines: Xlibris Publishing, 2007.

Maningas, Rad, *Balintawak: Lessons in Eskrima*. Bloomington, Indiana: iUniverse, 2015.

Miner, William D., LTC and Miner, Lewis A., *Surviving Hell: Surrender on Cebu*. New York, New York: Turner Publishing Co., 2010.

O'Reilly, Bill and Dugard, Martin, *Killing the Rising Sun: How American Vanquished World War II Japan*. New York, New York: Henry Holt and Company, LLC, 2016.

Russell, John, *The Balintawak System of Arnis-Eskrima*. Sudlon Publishing, 1994.

Segura, Manuel F. *Tabunan: The Untold Exploits of the Famed Cebu Guerrillas in World War II*. Cebu City, Philippines: MF Segura Publications, 1975.

Segura, Manuel F. *The Koga Papers: Stories of World War II*. Cebu City, Philippines: MF Segura Publications, 1992.

Warren, James Francis, *The Sulu Zone 1768-1898*. Singapore: National University of Singapore, 1981.

Wiley, Mark V. *Filipino Fighting Arts: Theory and Practice*. Burbank, California: Unique Publications, 2000.

About the Authors

Crispulo "Ising" Atillo

Crispulo "Ising" Atillo is the founder of Atillo Balintawak Eskrima – Original Saavedra Style and last remaining member of the Balintawak Self Defense Club founded in 1952. He has been featured in numerous international publications and produced several best-selling instructional DVDs on the art of Balintawak eskrima.

Crispulo Atillo maintains a hectic schedule teaching his beloved art of Atillo Balintawak Eskrima - Original Saavedra Style to students throughout the world at his training facility in Beaumont, California, and through his tremendously popular seminars throughout the United States.

For training or to schedule a seminar, contact Crispulo Atillo directly at (909) 363-6435, grandmasteratillo@yahoo.com, or www.atillobalintawak.com.

Glen Boodry

Glen Boodry is a recognized authority on the indigenous fighting arts of the Philippines, Indonesia, and Southeast Asia. He has trained extensively throughout the Philippines, Indonesia, Thailand, Japan, Africa, South America, and the Middle East. He is currently based in Idaho and continues to travel and research the martial arts all over the world.

www.ingramcontent.com/pod-product-compliance
Lightning Source LLC
Chambersburg PA
CBHW081209230426
43666CB00015B/2690